Catalysis for Sustainability

Goals, Challenges, and Impacts

Sustainability: Contributions through Science and Technology

Series Editor: Michael C. Cann, Ph.D.
Professor of Chemistry and Co-Director of Environmental Science
University of Scranton, Pennsylvania

Preface to the Series

Sustainability is rapidly moving from the wings to center stage. Overconsumption of non-renewable and renewable resources, as well as the concomitant production of waste has brought the world to a crossroads. Green chemistry, along with other green sciences technologies, must play a leading role in bringing about a sustainable society. The **Sustainability: Contributions through Science and Technology** series focuses on the role science can play in developing technologies that lessen our environmental impact. This highly interdisciplinary series discusses significant and timely topics ranging from energy research to the implementation of sustainable technologies. Our intention is for scientists from a variety of disciplines to provide contributions that recognize how the development of green technologies affects the triple bottom line (society, economic, and environment). The series will be of interest to academics, researchers, professionals, business leaders, policy makers, and students, as well as individuals who want to know the basics of the science and technology of sustainability.

Michael C. Cann

Published Titles

Green Chemistry for Environmental Sustainability
Edited by Sanjay Kumar Sharma, Ackmez Mudhoo, 2010

Microwave Heating as a Tool for Sustainable Chemistry
Edited by Nicholas E. Leadbeater, 2010

Green Organic Chemistry in Lecture and Laboratory
Edited by Andrew P. Dicks, 2011

A Novel Green Treatment for Textiles:
Plasma Treatment as a Sustainable Technology
C. W. Kan, 2014

Environmentally Friendly Syntheses Using Ionic Liquids
Edited by Jairton Dupont, Toshiyuki Itoh, Pedro Lozano, Sanjay V. Malhotra, 2015

Catalysis for Sustainability: Goals, Challenges, and Impacts
Edited by Thomas P. Umile, 2015

Sustainability: Contributions through Science and Technology

Series Editor: Michael C. Cann

Catalysis for Sustainability

Goals, Challenges, and Impacts

Edited by
Thomas P. Umile
Gwynedd Mercy University
Pennsylvania, USA

CRC Press
Taylor & Francis Group
Boca Raton London New York

CRC Press is an imprint of the
Taylor & Francis Group, an **informa** business

CRC Press
Taylor & Francis Group
6000 Broken Sound Parkway NW, Suite 300
Boca Raton, FL 33487-2742

ISBN 13: 978-1-4665-8685-7 (hbk)

Library of Congress Cataloging-in-Publication Data

Catalysis for sustainability : goals, challenges, and impacts / editor, Thomas P. Umile.
pages cm
Includes bibliographical references and index.
ISBN 978-1-4665-8685-7 (hardcover : alk. paper) 1. Catalysts. 2. Green chemistry. I. Umile, Thomas P., editor.

TP159.C3.C35 2015
660'.2995--dc23 2015027740

Visit the Taylor & Francis Web site at
http://www.taylorandfrancis.com

and the CRC Press Web site at
http://www.crcpress.com

Contents

Preface

How can we ensure that the growth and success of our society is not at the expense of future generations' prosperity? We must "simply" temper our aspirations with the recognition that a finite number of resources (natural, economic, or otherwise) are available to adequately meet our needs. Additionally, it helps to recognize that the decisions each person, community, and organization make affect each other and are interconnected. Achieving a sustainable society, accordingly, is a multifaceted challenge that requires input from entrepreneurs, governmental agencies and leaders, and the general public as well as scientific advancements and technological innovations. Providing goods and services in a sustainable manner requires the efficient manipulation of matter and energy; therefore, much responsibility falls upon chemists and chemical engineers to address the concerns of sustainability.

Catalysts are one of the many tools that the chemist has available to address the needs of sustainability. Catalysts, materials that enhance the speed of chemical reactions, have already found widespread application for the production of a vast majority of the products we consume, including food, clothing, pharmaceuticals, fuels, and plastics. Additionally, catalysts have an almost intrinsic capacity to directly address the needs of sustainability by lowering energy costs, making chemical reactions more selective, increasing the efficiency of resource consumption, and providing novel pathways for using alternative chemical feedstocks to produce goods.

Catalysis for Sustainability explores the intersection between catalytic science and sustainable technologies. Throughout this volume, the goals of research to develop new catalysts, the challenges to overcome, and the effects of success are discussed as they pertain to sustainability. The opening chapter provides an introduction to the science of catalysis and its potential effects on sustainability so that all readers are similarly oriented before encountering subsequent chapters that explore individual research areas in catalysis and sustainable science. In assembling a team of contributors, scientists were recruited from both academia and industry to provide diverse perspectives on research and practical applications. Their contributions explore various research areas within catalysis pertaining to sustainability, providing insights on the goals of a given field, the challenges to attaining those goals, and the (demonstrated or potential) effects of success. What follows are not comprehensive literature reviews. Each topic has been reviewed before and reviewed well, and a fitting, thorough treatment of each could fill entire volumes let alone a single chapter. Instead, our contributors explain the importance of a given field, describe how advances can help us to achieve a more sustainable world, present challenges to attaining such goals, and provide information on the effect of success.

A background in chemistry is helpful for fully navigating this text, but all readers should find something interesting herein. The "chemistry-heavy" sections can be skimmed in many places without forfeiting the narrative on the breadth, strategies, and intended impacts of catalysis research and development as they relate to sustainability. At its core, however, *Catalysis for Sustainability* is the book I wish

I had as a senior undergraduate or young graduate student. Although many excellent texts, review articles, and themed journal issues have been dedicated to the theme of catalysis and green or sustainable chemistry, such detailed reviews and technical accounts can be intimidating to a novice. (A thorough review article I recently came across "weighed in" at over 130 pages: an intimidating read for even an expert!) We attempted to make *Catalysis for Sustainability* accessible to as broad an audience as possible, but the young chemists and scientists out there are our primary audience. Our future scientists will inherit the problems of resource and energy use that older generations either created (or inherited ourselves without fixing). If this text inspires just one young, future scientist to focus their graduate research or career on the themes presented herein, I will consider this endeavor an overwhelming success.

Before I close, I must thank a number of people without whom the task of completing this text would have been impossible. Jack Kruper, Geoff Coates, Deana Zubris, and Mark Jones provided many stimulating conversations about catalysis and sustainability that influenced the shape of this project. Many thanks are necessary for a panel of volunteers who reviewed portions of this book and offered important advice as it developed: Laura E. Ator, Jacob W. Black, Nicholas C. Boaz, Elizabeth A. Burzynski, Robert W. Davis, Kimberly S. Graves, Melissa C. Grenier, Jyoti R. Tibrewala, Joseph C. Ulichny, and Purav P. Vagadia. Hilary Rowe, and Laurie Oknowsky at Taylor & Francis deserve special recognition for dealing with all of the delays and "11th hour" requests that accompanied the preparation of our final product. Finally, I am infinitely grateful to series editor Michael C. Cann for the invitation to prepare this volume and for first introducing me to green chemistry during a lively 9:00 a.m. organic chemistry lecture many years ago.

Thomas P. Umile
Gwynedd Valley, Pennsylvania

Editor

Thomas P. Umile joined the Division of Natural and Computational Sciences at Gwynedd Mercy University in 2014 as assistant professor of chemistry. He completed his undergraduate studies at The University of Scranton in 2006, where he studied microwave-assisted organic reactions and green chemistry with Michael C. Cann. Afterward, he joined the laboratory of John T. Groves at Princeton University, earning his PhD in 2012 for the development of chlorine dioxide-generating metalloporphyrin catalysts. Prior to his faculty appointment, Tom was a postdoctoral fellow at Villanova University, where he taught organic chemistry and worked with Kevin P. C. Minbiole to isolate and characterize bioactive small molecules from microbial sources. His current academic interests include the isolation and characterization of natural products and infusing undergraduate courses with green chemistry and sustainability.

Contributors

Jeremy R. Andreatta
Department of Chemistry
Worcester State University
Worcester, Massachusetts

Ryan S. Buzdygon
HepatoChem, Inc.
Beverly, Massachusetts

Meghna Dilip
Department of Chemistry
Worcester State University
Worcester, Massachusetts

Matthew R. Dintzner
College of Pharmacy
Western New England University
Springfield, Massachusetts

Bryan C. Eigenbrodt
Department of Chemistry
Villanova University
Villanova, Pennsylvania

Craig Jamieson
Department of Pure and Applied Chemistry
University of Strathclyde
Glasgow, Scotland, United Kingdom

Birgit Kosjek
Merck Research Laboratories
Rahway, New Jersey

Erika M. Milczek
Merck Research Laboratories
Rahway, New Jersey

Philip Nuss
Center for Industrial Ecology
School of Forestry and Environmental Studies
Yale University
New Haven, Connecticut

Jared J. Paul
Department of Chemistry
Villanova University
Villanova, Pennsylvania

Margaret H. Roeder
Department of Chemistry
Villanova University
Villanova, Pennsylvania

Thomas P. Umile
Division of Natural and Computational Sciences
Gwynedd Mercy University
Gwynedd Valley, Pennsylvania

Allan J. B. Watson
Department of Pure and Applied Chemistry
University of Strathclyde
Glasgow, Scotland, United Kingdom

1 Catalysts and Sustainability

Thomas P. Umile

CONTENTS

1.1 INTRODUCTION

Catalysts are compounds that increase the rate of chemical reactions without themselves being consumed in the process. Accordingly, this enhancement makes them attractive for the manufacture of chemical products. Because most of the goods and energy sources we use require such chemical processing, catalysts are fundamental in our daily lives. Additionally, enzymes are nature's biological catalysts, making the biochemical transformations necessary for life possible. Indeed, catalysts have a consistent and broad effect on our daily lives.

Beyond simply "speeding up a chemical reaction," catalysts have an intrinsic potential to address our concerns and needs for a sustainable human existence. Because the world's population is exponentially growing, we are faced with our civilization's greatest challenge: to satisfy the needs of our current way of life without jeopardizing our ability to do the same tomorrow. Catalysts' chemical rate enhancement means that they can contribute to overcoming this challenge by lowering the energy costs of chemical reactions, making such reactions more selective and less wasteful, and offering opportunities for developing new chemical reactions that make efficient use of consumed materials. Thus, catalysts can have dramatic, beneficial effects on the use of environmental resources, the development of new products and technological solutions, and the financial concerns of people, businesses, and manufacturers. In this chapter, we will explore the needs of a sustainable society with special respect to the responsibilities of science, technology, and especially catalysis, review the fundamentals of catalytic chemistry and how such processes are evaluated, and provide introductory examples in which catalysts have already begun to affect our world to make it more sustainable.

1.2 SUSTAINABILITY AND GREEN CHEMISTRY

In 1987, the Brundtland Commission released *Our Common Future*, a report that detailed the group's findings related to issues about the environment and world development. In it, the commission defined *sustainable development* as "meeting the needs of the present without compromising the ability of future generations to meet their own needs."[1] These candid, pragmatic words describe the only way that humans can achieve an existence that is both productive and long-lived.

One measure to which sustainability is often held is the "Triple Bottom Line," a term coined by business consultant John Elkington.[2] The Triple Bottom Line reminds us of our responsibilities to three stakeholders: the environment, the economy, and society. These have sometimes been referred to as the "Three P's" of people, planet, and profit.[3] The needs of the Three P's must each be adequately addressed for a product, service, or practice to be truly sustainable. Although the Triple Bottom Line may have originally been conceived as a corporate philosophy or guide, scientists and researchers must also consider environmental, economic, and social needs when developing new technologies.

The environment, perhaps most commonly associated with themes of sustainability, is our finite storehouse of resources that are used to create, manufacture, and power all the goods that humans use. These include not only basic needs such as food, clothing, and shelter but also all the consumables and goods that are a part of daily life. Pharmaceuticals and healthcare equipment, automobiles, televisions, and smartphones all are produced using the planet's resources. Moreover, the power necessary to process and cook food, drive cars, and use electronic devices all come from the planet's resources. Unfortunately, many of these resources are nonrenewable.

Petroleum has traditionally been our primary feedstock for all organic (carbon containing) compounds necessary to manufacture goods.[4] Useful compounds from petroleum include aromatic compounds such as benzene and olefins such as ethylene, as well as synthesis gas (a mixture of carbon monoxide and hydrogen).[5] These petrochemicals are not used as fuel but as the raw material for synthesizing products including pharmaceuticals, plastics and resins, solvents, lubricants, dyes, cosmetics, and apparel. In 2013, approximately 15% of all petroleum refinery yield in the United States was for nonfuel uses.[6] Simply put, just about anything that is not composed of a mineral, metal, plant, or other living organism is made from petroleum.[7]

A sustainable approach to the environment recognizes the irrefutable fact that if we consume a resource (such as petroleum) faster than it can be replenished, *it will eventually run out*. One can debate how long it will take to run out, but the eventual depletion of a finite resource is logically unavoidable. Because the planet's resources are crucial to our existence, it is necessary to properly care for them. Importantly, these resources include not only raw materials (e.g., minerals, air, and water) for goods and fuels but also the complex, biodiverse, and interconnected ecosystems that provide services beneficial to human existence including pollination, nutrient cycling, and climate regulation.[8,9]

The Triple Bottom Line also recognizes the importance of economic factors to a developing, sustainable world. The economy provides the financial capital that drives business, governments, and trade, funds the development of technological advances, and supports our own personal activities. Although a discussion of the principles for a sustainable business is beyond our scope here, the scientist or engineer must also consider the economic implications of their work. The financial costs to the consumer and industry to purchase, use, or implement a technology are real concerns. Focusing closely on environmental interests (e.g., avoiding pollution, reducing waste, and saving energy) is noble but will not translate into the development of a viable technology if the economic or financial effects are not simultaneously considered.

Sustainability's final responsibility, but by no means its least important, is to society. Technologies, policies, and corporations that offer solutions to environmental and economic problems must also meet the needs and demands of society to be considered truly sustainable. This is not to imply that sustainability is solely at the whims of society's values. Values change over time, and indeed evaluating our own lifestyles and personal needs is one strategy for promoting a sustainable society. However, the science and technology sector can have a positive effect on society by simply renewing the public's trust and understanding of science. Chemists today know that many nonscientists harbor a perhaps ironic distrust of "chemicals." (Ironic because everything is made of chemicals!) This "chemophobia"[10,11] is born out of a public misunderstanding of science[12] and a history of attention-grabbing news related to the toxicity of some compounds,[13] health risks from sources once championed as safe,[14] and pollution or environmental disasters resulting from technology (such as the Deepwater Horizon oil spill in 2010 and the Fukushima power plant meltdown in 2011). Developing technologies that minimize or completely avoid such risks would do a great service to society, and safer products could contribute to a stronger relationship between scientists and the general public.

The Triple Bottom Line reminds us that a sustainable technology, policy, or practice is one that must meet the needs of the environment, economy, and society *simultaneously*. A brighter, energy-efficient lightbulb is no solution if it is so expensive that few can purchase it. A cheaper, more energy-efficient lightbulb is no solution if it does not burn as brightly as the user demands. A cheaper, brighter lightbulb is no solution if it consumes more energy. Simply put, the lightbulb needs to be all three: brighter, cheaper, *and* energy-efficient.

Importantly, the three elements of the Triple Bottom Line do not exist in a vacuum. Perturb one, and the others immediately respond. Fortuitously, positive changes to one facet can have desirable effects elsewhere.[15] For example, consider that some new technology is developed to manufacture a product while eliminating the generation of a hazardous waste. Less waste translates to less garbage to dispose of, less pollution, and a more efficient consumption of resources (environment wins). In this era of (necessary) environmental regulations and policies, less waste generally means reduced regulatory oversight, fewer disposal costs, and a reduced risk of potential fines. Furthermore, a process that uses resources more efficiently can reduce manufacturing costs. These financial savings strengthen the corporation and are perhaps partly passed onto the consumer (economy wins). Finally, fewer wastes mean that landfills fill up more slowly, and the public is not at risk for exposure to potentially hazardous wastes (society wins). In fact, these beneficial changes could have downstream consequences, further reinforcing the positive changes. The cheaper product sells more and, in our environmentally aware society, the cleaner technology could create good will that fosters customer loyalty. Simply put, the implementation of truly sustainable technologies can have many favorable, long-term effects.

In 1998, Paul Anastas and John Warner published *Green Chemistry: Theory and Practice*, which called on chemists to design (and redesign) chemical processes that reduce or completely eliminate the generation of hazardous wastes.[4] The development of green chemistry was motivated primarily by two factors: prominent environmental disasters (such as the Cuyahoga River fire in Ohio and the contamination of the Love Canal neighborhood in upstate New York) and a growing regulatory oversight of pollution. Green chemistry offered an alternative philosophy to so-called "end of pipe" solutions to waste. Rather than collect, clean, or otherwise neutralize pollution once it is generated, why not avoid its generation altogether? Chemistry is fundamentally about the molecular basis of matter and the changes that occur during chemical reactions; therefore, much of the responsibility for the efficient use of matter falls on chemists. Anastas and Warner thus developed the Twelve Principles of Green Chemistry (see next page) as guidelines to assist chemists in creating "environmentally benign" reactions and processes that reduce or outright avoid the generation of waste.

The word "sustainability" appears on just two pages of that original book,[4] yet green chemistry has since become synonymous with "sustainable chemistry" in many uses. Although some draw a distinction between the two,[16] if one examines the Twelve Principles of Green Chemistry through the lens of the Triple Bottom Line, it becomes apparent that the original tenets of green chemistry actually align quite well with the goals of sustainability. Environmental, economic, and social concerns are explicitly addressed throughout green chemistry. Green chemistry and sustainability are inextricably linked. One might say that green chemistry is the method by which the practicing chemist approaches sustainability. (However, we shall eventually see that a more holistic view, beyond just chemistry, is critically needed to be truly sustainable!)

The Presidential Green Chemistry Challenge Awards were initiated in 1996 to celebrate the development and implementation of new technologies that exemplify the values of green chemistry.[17] The awards are presented annually by the US Environmental Protection Agency. Notably, these technologies are not merely conceptual but often demonstrated through successful practice in the marketplace. The public comes into contact with the results of Presidential Green Chemistry Challenge Award–winning technologies in everyday life. Examples of such technologies include NatureWorks polylactic acid[18] (a compostable plastic produced from fermented plant biomass and used to make products such as disposable beverage containers and produce packaging) and Sherwin-Williams' water-based acrylic paint technology that reduces the emission of volatile organic compounds.[19] In addition to addressing the concerns of green chemistry, these technologies demonstrate utility

THE TWELVE PRINCIPLES OF GREEN CHEMISTRY

1. It is better to prevent waste than clean up waste after it is formed.
2. Synthetic methods should be designed to maximize the incorporation of all materials used in the process into the final product.
3. Wherever practicable, synthetic methodologies should be designed to use and generate substances that possess little or no toxicity to human health and the environment.
4. Chemical products should be designed to preserve efficacy of function while reducing toxicity.
5. The use of auxiliary substances (e.g., solvents and separation agents) should be made unnecessary wherever possible and innocuous when used.
6. Energy requirements should be recognized for their environmental and economic effects and should be minimized.
7. A raw material or feedstock should be renewable rather than depleting wherever technically and economically practicable.
8. Unnecessary derivatization (blocking group, protection/deprotection, temporary modification of physical/chemical process) should be avoided wherever possible.
9. Catalytic reagents (as selective as possible) are superior to stoichiometric reagents.
10. Chemical products should be designed so that at the end of their function they do not persist in the environment and break down into innocuous degradation products.
11. Analytical methodologies need to be further developed to allow for real-time, in-process monitoring and control before the formation of hazardous substances.
12. Substances and the form of a substance used in a chemical process should be chosen so as to minimize the potential for chemical accidents, including releases, explosions, and fires.

Source: Anastas, P.T., and Warner, J.C. *Green Chemistry: Theory and Practice* 1998. Reproduced with permission of Oxford University Press.

to the public and commercial success in the marketplace, and so these technologies indeed address the concerns of the Triple Bottom Line. It is important to note that although the Presidential Green Chemistry Challenge Awards bring much-needed attention to environmentally benign technologies, they are not the *only* instances of green technologies that have reached industrial practice. As business leaders recognize the benefits sustainability can have on their own bottom line (in addition to the Triple Bottom Line), more and more companies are incorporating such technologies even without the public distinction.

1.3 CATALYSIS AND SUSTAINABILITY

Of particular interest to this text, one principle of green chemistry states that selective catalysts are desirable for chemical reactions.[4] A catalyst is a substance that increases the rate of a chemical reaction while not being consumed during that reaction.[20] Catalysts earned such distinction among the Twelve Principles of Green Chemistry because of the almost intrinsic, positive effects that catalysts can have on creating safer, less-wasteful reactions. Catalysts can lower the energy costs of a chemical transformation, increase the speed at which a product is generated, diminish the occurrence of unproductive or wasteful side reactions, and enable the practical use of previously untapped precursors or reactions. Perhaps most importantly, because they are unchanged during the course of the reaction they enhance, catalysts are theoretically reusable.

Catalysts are ubiquitous in our everyday lives. More than 90% of industrial chemical processes employ catalysts,[21] implying that most of the goods we use are produced using catalysts. Many of us come in close contact with a catalyst daily when driving our automobiles. Automobiles contain catalytic converters that facilitate the conversion of highly undesirable combustion products (e.g., nitrogen oxides, carbon monoxide, and unreacted hydrocarbons) into less hazardous gases.[22] Enzymes are also catalysts, enhancing the speed and selectivity of important biochemical reactions in all living systems. Enzymes are responsible for everything from the synthesis of complex biomolecules to the simple, but quite necessary, oxidation of water to produce the oxygen we breathe.

Of the more than 100 technologies that have been honored with Presidential Green Chemistry Challenge Awards, no fewer than 14 are *specifically* for a novel catalytic method, and many more use catalysis at some point in their overall implementation.[17] Despite the potential and *proven* benefits of catalysis, however, their use and realization can be complicated. Before being implemented, fast, selective, and long-lived catalysts must first be developed, and the constraints of sustainability further challenge researchers. Historically, many well-behaved and understood catalysts employ expensive, toxic, or rare precious metals, in direct contrast with the needs of the Triple Bottom Line. Additionally, replacing a time-tested, large-scale industrial process with a newly developed, sustainable alternative can require prohibitive capital costs that are not financially viable in the short-term. Regardless, catalysis remains an important goal for developing a more sustainable world, and much work is being done to both refine current catalytic processes to meet the demands of the Triple Bottom Line and to develop new catalysts to address the sustainability demands of the twenty-first century.

However, can sustainable catalysts have an effect at the global proportions necessary? In addition to the ubiquity of catalyst technologies discussed, it is worth considering that one single catalytic technology has already had an effect of the magnitude we need to favorably affect sustainability. The Haber–Bosch process is a method for converting nitrogen (N_2) and hydrogen gas (H_2) into ammonia (NH_3) using an iron catalyst. It was invented by Fritz Haber and later implemented by Carl Bosch in 1913.[23] (The two would later win separate Nobel Prizes related to their work on this process.) The Haber–Bosch process has a storied past, as ammonia produced in this way was, and still is, used heavily for the production of military explosives.[24]

Today, however, the process is primarily used toward a far more prosperous end: the production of fertilizers.[25] Worldwide production of ammonia was as high as 140 megatons in 2013,[26] and the global population of nearly 7 billion people is supported in large part by Haber–Bosch ammonia.[27] Estimates suggest that fertilizers produced from Haber–Bosch nitrogen are used to feed up to 48% of the world,[24] and the Haber–Bosch process has been called the "detonator of the population explosion" that has occurred over the last 100 years.[28] Indeed, one catalytic technology can have quite the global effect.

1.4 HOW CATALYSTS WORK

To appreciate the exploration of catalyst design and implementation that is to follow, a brief overview of the fundamental chemistry of catalysis is presented below. Entire textbooks have been written on the subject,[29,30] and by no means is the following a replacement for any of them. The interested reader is directed to such references for a more detailed discussion.[29–31]

During a chemical reaction, the reacting system undergoes energy changes related to bond-making and bond-breaking (enthalpy) and the disorder of the system (entropy). Together, these factors contribute to the *thermodynamics* of the reaction. Reactants and products can exist at different potential energy levels that dictate their propensity to react (just as a bowling ball at the top of a hill has more potential energy to roll away than one sitting at the bottom of the same hill). The difference in potential energy between the reactants and products is oftentimes called the *driving force* (ΔE)* of a reaction (Figure 1.1), and thermodynamics tracks the energy changes that occur during a reaction because of this difference. If the reactants of a chemical reaction are higher in energy than the products, the reaction is said to be energetically *downhill* (or "exergonic") and occurs spontaneously. Examples of spontaneous chemical processes include paper burning, an iron nail rusting, and diamonds turning into graphite. As should be obvious from this list, not all energetically favorable, spontaneous reactions are fast (e.g., iron nails rusting), some might even be slow enough that we need not worry about them during our lifetimes (e.g., diamonds turning into graphite), and others may need a "kick," so to speak, to get going (e.g., a spark to ignite a paper fire). (Hamori and Muldrey suggested that the more anthropomorphic word "eager," which does not have speed connotations, better describes the tendency for compounds in an exergonic reaction to react rather than "spontaneous.")[32]

The differences in rate at which chemical reactions proceed arise from the second major factor that mediates chemical reactivity: the activation energy. When a chemical reaction occurs, the atoms and bonds in the molecules of the reactants rearrange to produce the products. This rearrangement requires an input of energy, called the *activation energy* (E_{act}; Figure 1.1). The activation energy is an energetic barrier that dictates the rate at which a chemical reaction occurs. A larger activation energy corresponds to a slower reaction. (It is perhaps not incorrect to conceptualize E_{act}

* What is being called driving force (ΔE) here is, more accurately, known as Gibbs free energy (ΔG). The latter term is not introduced for clarity in this section.

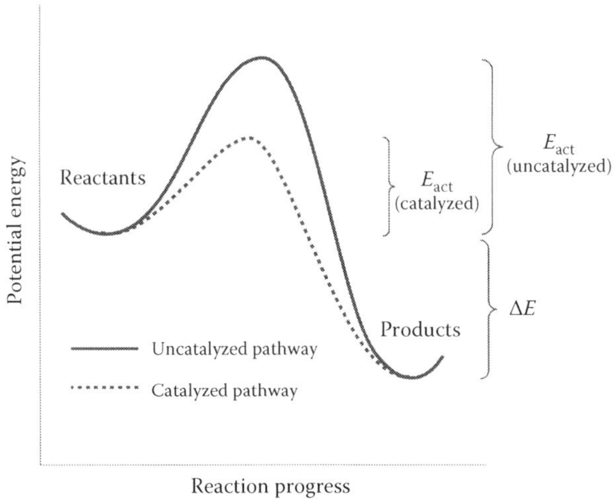

FIGURE 1.1 Representative reaction coordinate diagram for a catalyzed and uncatalyzed exergonic reaction.

as a mountain over which the reactants must traverse to become products. Smaller mountains can be crossed faster than larger ones.)

A catalyst enhances the rate of a chemical reaction by lowering the reaction's E_{act} (Figure 1.1). The relationship between activation energy and reaction rate is exponential, implying that even modest reductions in the activation energy for a reaction can result in dramatic increases in reaction rate. It is important to note, however, that a catalyst does not change the driving force of the reaction. A catalyst will never make a nonspontaneous reaction spontaneous.

Chemical reactions are ultimately a combination or other rearrangement of the atoms in the reactants to form products. The precise, step-by-step process of bond-making and bond-breaking by which reactants become products is called its *mechanism*. A catalyst normally lowers the activation energy of the overall reaction by providing an alternate, lower-energy mechanism for a reaction to occur. Consider the simple example of alkene hydration (Figure 1.2). Under normal conditions, water and

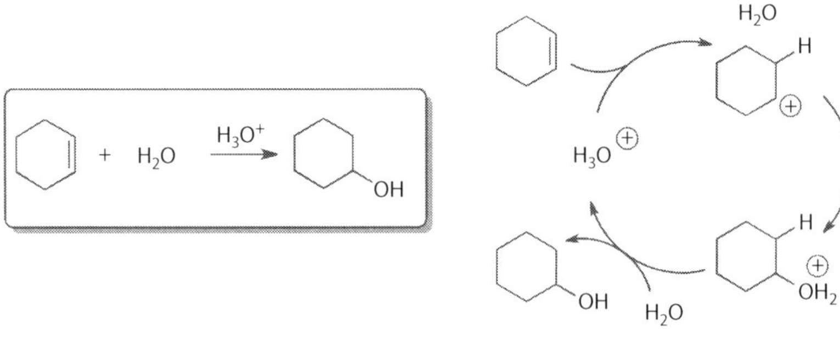

FIGURE 1.2 Acid-catalyzed hydration of cyclohexene and its mechanism.

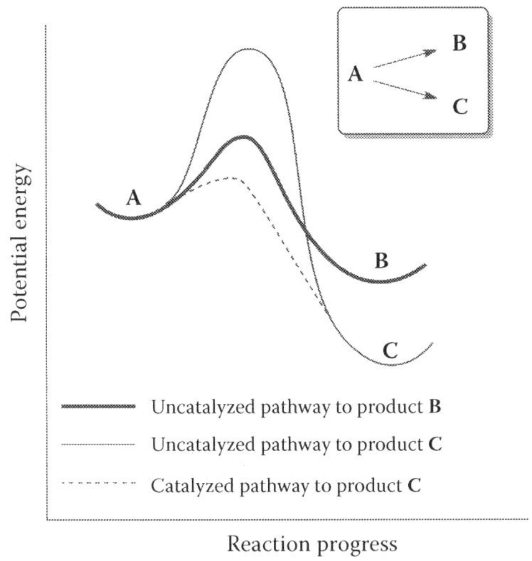

FIGURE 1.3 Example reaction coordinate diagram for two competing reactions (i.e., **A** → **B** and **A** → **C**), demonstrating how a catalyst can selectively enhance one (**A** → **C**).

alkenes do not appreciably react with one another. However, in the presence of aqueous acid (i.e., hydronium ion, H_3O^+), the reaction proceeds smoothly. Here, H_3O^+ first reacts with the alkene to generate a reactive carbocation intermediate and water. The carbocation and water combine to form a reactive alkyloxonium ion that loses a proton to another water molecule, producing the alcohol product and regenerating H_3O^+ in the process. H_3O^+, because it is not consumed during the reaction, is a catalyst. Moreover, in the absence of H_3O^+, the first step of the shown mechanism would be impossible, and some other, higher energy (i.e., slower) mechanism would be necessary.

The property of a catalyst to reduce a reaction's activation energy can also assist in controlling that reaction, particularly when the potential exists for multiple, different products. If, for example, reactant **A** is capable of producing one of two possible products, **B** or **C**, the catalyst can influence which product is produced. Under so-called "kinetic conditions" (e.g., lower temperatures), the product produced is normally the product produced by the pathway with lower activation energy. In Figure 1.3, the conversion of **A** → **B** would occur faster than the conversion of **A** → **C** because the pathway to **B** has a lower activation energy. Thus, the major (or even sole) product of the reaction would be **B**. If, however, a catalyst was designed and used that selectively lowered the activation energy leading to product **C** (dashed line in the figure), product **C** would instead be produced faster under the same conditions. The catalyst here provides a lower-energy, alternative pathway for **C** alone and thus would be considered a selective catalyst.

1.4.1 Types of Catalysts

Traditionally, catalysts can be divided into two classes depending on their physical state and the state of the reactants upon which they act. A *homogeneous catalyst* is in the same physical state as the reactants, which usually means that the catalyst is dissolved in the same solvent as the reactants. Common examples range from simple acids and metal ions to more intricate inorganic and organometallic complexes. A *heterogeneous catalyst*, by contrast, is in a different physical state than the reactants. Heterogeneous catalysts are frequently insoluble solids such as metals or clays that catalyze reactions in the liquids or gases around them.

From a practical, industrial perspective, heterogeneous catalysts offer a great advantage because they can be easily separated when the desired reaction is complete. The catalyst can be simply removed by some physical method such as filtration or centrifugation. Indeed, most industrial catalysts are heterogeneous.[33] Homogeneous catalysts, dissolved and dispersed in the reaction medium, are often much harder to remove and recover after the reaction is complete and, thus, are less desirable for large-scale operations.

Difficulty in separation aside, however, homogeneous catalysts are not without their own benefits. Precisely because they are dissolved and dispersed directly in the reaction medium, homogeneous catalysts are more apt to encounter reactant molecules. This typically results in higher activities than heterogeneous catalysts, and most homogeneous catalysts operate at milder reaction conditions (e.g., <200°C).[34] Moreover, because they have discrete and well-defined molecular structures, the precise mechanism by which a homogeneous catalyst operates is usually better understood.[34] An understanding of the chemical mechanisms dramatically facilitates the development of better, faster, and more selective catalysts. To capitalize on the benefits of both heterogeneous and homogeneous catalysts, some developed homogeneous catalysts have successfully been made heterogeneous by supporting them on an insoluble material such as a polymer matrix or clay.

For reactions that require a biphasic system, a *phase transfer catalyst* can be used. Some reaction components exist in two different, immiscible phases. For instance, one may desire to react a water-soluble reagent with an organic, hydrophobic reagent. To facilitate the interaction of these two substances, a compound is added that is capable of noncovalently binding to one of the two reagents and bringing it into the other phase (Figure 1.4). Accordingly, a phase transfer catalyst for this purpose

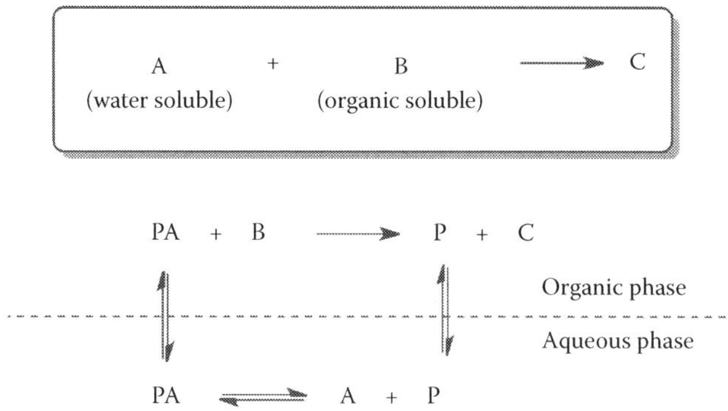

FIGURE 1.4 A phase transfer catalyst (P) allows two reagents in different phases (A and B) to interact by enhancing the solubility of one reagent in the other phase.

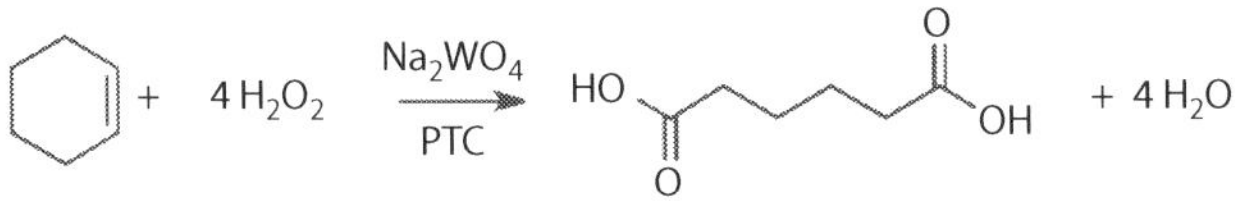

FIGURE 1.5 Alternative synthesis of adipic acid using hydrogen peroxide and a phase transfer catalyst (PTC), which avoids the generation of nitrogen oxide emissions.[35]

has both hydrophobic and hydrophilic components. Typical phase transfer catalysts include crown ethers and quaternary ammonium salts.[22] Phase transfer catalysts not only facilitate the dissolution of a reagent in another phase but also enhance reactions because ionic reagents (i.e., the water-soluble component) are usually more reactive in the organic phase. Additionally, the use of two separate phases can make the separation of the product from the reaction mixture easier (e.g., if the product is soluble in one phase but not the other).[22]

An example of a phase transfer catalyst "in action" is Noyori's synthesis of adipic acid from cyclohexene (Figure 1.5).[35] Adipic acid, a precursor for making nylon, is commonly prepared using concentrated nitric acid to oxidize a mixture of cyclohexene and cyclohexanol.[36] This method generates nitrogen oxide emissions as byproducts. NO and NO_2 produced in this way can be collected and used to produce nitric acid (effectively recycled), but the potent greenhouse gas N_2O cannot be similarly recycled and must instead be treated as pollution.[37] Noyori's alternative avoids the generation of nitrogen oxide wastes altogether by using hydrogen peroxide (H_2O_2) to oxidize cyclohexene, catalyzed by sodium tungstate (Na_2WO_4).[35] Hydrogen peroxide oxidizes the tungstate (in the aqueous phase), which in turn oxidizes cyclohexene (in the nonpolar, organic phase). A quaternary ammonium phase transfer catalyst increases the solubility of the oxidized tungstate in the organic phase, facilitating its encounter with and oxidation of cyclohexene.

1.5 ASSESSING THE EFFICACY OF CATALYSTS AND REACTIONS

To determine how well a catalyst or chemical reaction meets our needs, a number of common metrics or other quantitative tools are commonly used. These allow us to assess different facets of the reaction or process of interest, with respect to both its intrinsic success for producing the desired chemical product as well as its effect with regard to green chemistry and sustainability. The effectiveness of a catalyst can be quantified with regard to its activity, speed, reusability, and selectivity. Other metrics allow one to determine the effect, particularly waste generation, of a chemical process. Together, all of these factors can affect the Triple Bottom Line.

A catalytic turnover is described as one "turn" of the catalytic cycle. Recall that during a catalyzed reaction, the catalyst provides a lower-energy mechanism that eventually leads to both the production of product and the regeneration of the catalyst. That process, from reactant to product with regeneration of the initial catalyst species, is one "turn" of the catalytic cycle or one turnover. Two values are commonly reported for assessing the efficacy of a catalyst based on the concept of the turnover: *turnover number* and *turnover frequency*.

The *turnover number* (TON) is the total number of turnovers that a single catalyst molecule or site can perform, and a single turnover normally produces one molecule of product. For a homogeneous catalyst, TON is calculated as

$$\text{Turnover number (TON)} = \frac{\text{Moles of product}}{\text{Moles of catalyst}}$$

For a heterogeneous catalyst, the calculation requires one to know the number of active catalyst sites each insoluble particle makes available. TON essentially tells us how much product a catalyst can produce. In theory, a catalyst's TON should equal infinity because the catalyst is always regenerated during a turnover. In reality, a catalyst can become deactivated during multiple turnovers. For example, side reactions can damage an organometallic or inorganic complex's delicate ligands, or the catalyst can simply decompose. Additionally, by-products or impurities can poison the catalyst by reacting with it in an undesired way and deactivate it. For industrial applications, TONs of 10^6 to 10^7 are desirable.[38]

A particularly dramatic example of a "catalyst" with a low TON is the Friedel–Crafts acylation of aromatic compounds (Figure 1.6). This reaction, commonly taught in undergraduate classrooms and laboratories, uses an aluminum chloride "catalyst" to activate an acyl chloride that then reacts with an aromatic molecule. Although the mechanism of this reaction, on paper, suggests that $AlCl_3$ is not consumed by the process, the aryl ketone product forms complexes with $AlCl_3$ that prevent the $AlCl_3$ from further catalyzing any subsequent reactions.[39] Therefore, in practice, a complete, *stoichiometric* amount of $AlCl_3$ is necessary. Essentially, $AlCl_3$ only completes one turnover before it is deactivated.

Related to TON is the concept of *catalyst loading*, which is the number of moles of catalyst used in a reaction relative to the number of moles of (limiting) reactant.

$$\text{Catalyst loading} = \frac{\text{Moles catalyst}}{\text{Moles of limiting reactant}} \times 100\%$$

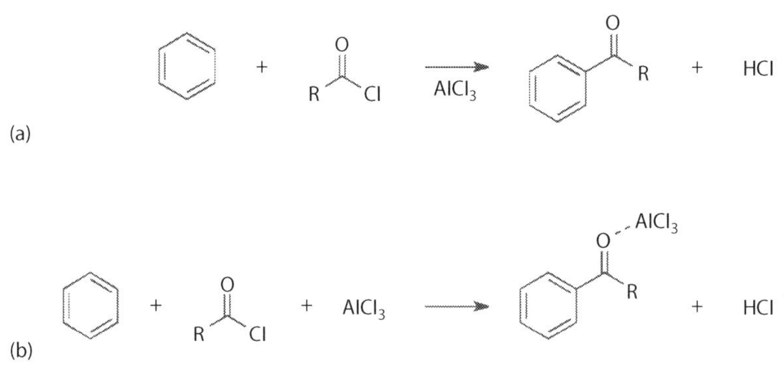

FIGURE 1.6 Two different ways of presenting the Friedel–Crafts acylation. (a) Emphasizing $AlCl_3$ as a catalyst. (b) Emphasizing that $AlCl_3$ is consumed by the reaction.

Higher catalyst loadings are required for those catalysts that do not have high turnover numbers, and one can normally infer a catalyst's turnover number from its loading. For example, a reaction that requires 20% catalyst loading (e.g., a 5:1 ratio of reactant to catalyst) normally means that the catalyst can only carry out five turnovers in the desired amount of time. Ideally, the catalyst loading would be kept as low as possible.

The number of turnovers that a single catalyst molecule or site can perform in a given amount of time is the *turnover frequency* (TOF). TOF provides information of rate, and its units are reciprocal time (e.g., s^{-1}, min^{-1}, h^{-1}). TOF is calculated by dividing TON by the total amount of time the reaction took to complete. For example, a catalyst that undergoes 1000 turnovers in 5 min has a TOF of 200 min^{-1} (i.e., 200 equivalents of product are produced per catalyst site per minute). Naturally, higher TOF is usually desirable.

When a reaction can conceivably produce a mixture of products, especially a mixture of stereoisomers, it is important to recognize the extent to which the catalyst favors the formation of one product over another. This selectivity is often reported as a ratio of the products to each other. For example, one method of preparing a key intermediate for synthesizing atorvastatin (Lipitor®) uses a ketoreductase enzyme to reduce ethyl-4-chloroacetoacetate to ethyl-4-chloro-3-hydroxybutyrate.[40] The alcohol product is chiral and could conceivably be a mixture of two isomers. However, ketoreductase generates the desired *S* enantiomer over the *R* enantiomer by a ratio of more than 399:1. (For enantiomers like this, the product ratio is sometimes reported as an *enantiomeric excess* [e.e.], which describes what percentage of the total product is the major isomer. Here, e.e. = 99.5%.) See Figure 1.7.

In addition to metrics for quantifying the efficacy of a catalyst with regard to its activity, speed, and selectivity, it is often desirable to consider how effective a catalyst (or indeed, any reaction) is with regard to the concerns of green chemistry.[41] Accordingly, metrics have been developed for evaluating such features. Jiménez-González et al.[41] have written an excellent primer on the use of "green" metrics in the pharmaceutical industry.

Atom economy is perhaps the most recognized metric invoked for assessing the potential for waste generation during a chemical reaction. The concept of atom economy evaluates the efficiency of a reaction by assessing how many of the atoms used to synthesize a product (by mass) actually end up in the desired product. Not all reagents used in a chemical reaction become incorporated into the product, and those that do not become wastes. For example, consider the textbook reduction of a ketone by sodium borohydride (Rxn. 1). Although this reaction is essentially the addition of two hydrogen atoms to the carbonyl π-bond, far more than just two hydrogen atoms are used per ketone molecule. Accordingly, by-products (i.e., wastes) are generated.

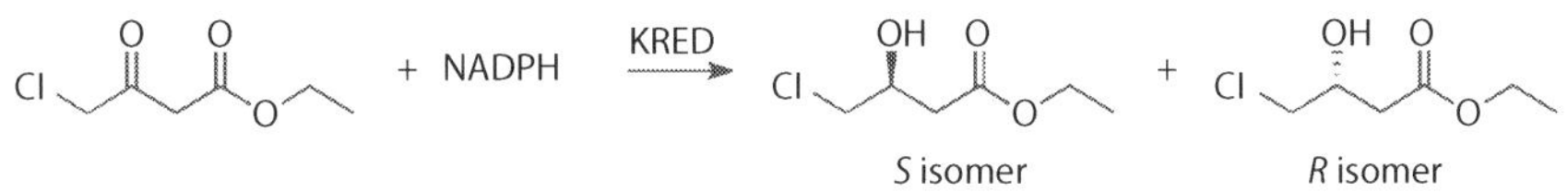

FIGURE 1.7 Enantioselectivity of ketoreductase (KRED) on the reduction of ketones to alcohols.[40] (NADPH: reduced nicotinamide adenine dinucleotide phosphate.)

A more atom economical reduction would use simply H_2 (Rxn. 2). In practice, however, most fine chemicals have more complex molecular structures than those in this example, and so increasingly complex reagents have been historically developed to provide increasingly greater levels of chemocontrol, regiocontrol, or stereocontrol to a reaction. Barry Trost recognized that the efficient use of all atoms employed in a chemical reaction was as important as these more common markers of reaction efficacy,[42] and atom economy became incorporated as a principle of green chemistry.[4]

(Rxn. 1)

(Rxn. 2)

Atom economy is reported quantitatively by dividing the formula weight of the desired product by the total formula weights (FW) of all the compounds used in the reaction.*

$$\text{Atom economy} = \frac{\text{FW of desired product}}{\text{FW of all reactants}} \times 100\%$$

For example, in the two reactions described above:

$$\text{Atom economy for Rxn. 1} = \frac{(60.1 \times 4)}{(58.1 \times 4) + 37.8 + (18.0 \times 4)} \times 100\% = 70\%$$

$$\text{Atom economy for Rxn. 2} = \frac{60.1}{58.1 + 2.0} \times 100\% = 100\%$$

Atom economy gives an interpretable, numerical value for what percentage of the atomic masses input to a chemical reaction end up in the final product. The alcohol product of Reaction 1 incorporates 70% (by mass) of all the atoms used in the reaction. The remainder of the mass input into the reaction (i.e., 30% of the mass) is not product. Reaction 2, comparatively, uses all of the atoms input to the reaction, and thus no waste is generated. Higher atom economy values are, accordingly, desirable.

Catalysts can often be useful for improving a process to maximize atom economy. In the simple ketone reduction example above, the direct hydrogenation (Reaction 2)

* One could also calculate atom economy by dividing by the total formula weight of *all products and wastes* in a reaction because of mass balance (FW of all reactants = FW of desired product + FW of waste).

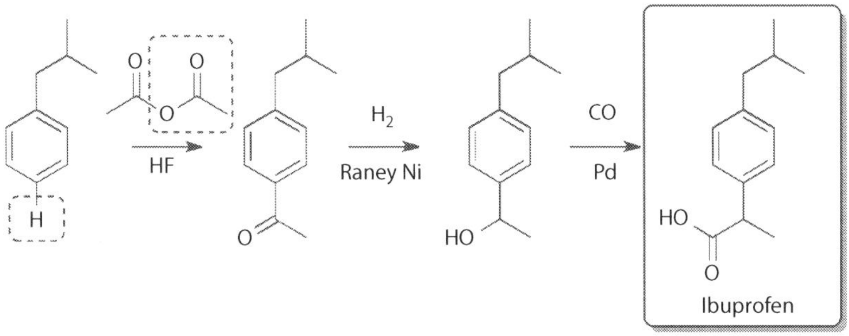

FIGURE 1.8 BHC synthesis of ibuprofen. Unused atoms are circled with a dashed line. Reused catalysts are bolded and not included in the calculation.

is only accessible through the use of a catalyst* (commonly based on a metal such as platinum, palladium, or rhodium). Trost, in a follow-up to his introduction of atom economy, noted that homogeneous catalysts "lead the way" in improving atom economy.[43] A particularly noteworthy example where catalysts have improved atom economy is the improved synthesis of ibuprofen, introduced by the BHC Company (now BASF) in 1992 (Figure 1.8).[44,45] The details of its synthesis and implementation have been described and reported in case studies[46,47] elsewhere as a triumph of green chemistry. Simply put here, however, this revised synthesis uses three catalytic steps to prepare ibuprofen from isopropyl benzene with an atom economy of 77%. The previously practiced method, known as the Boots synthesis, required six steps and had an atom economy of 40% (Figure 1.9).[46,47] (Note the use of the stoichiometric "catalyst" $AlCl_3$ in the Boots synthesis.)

Atom economy indicates if a reaction will intrinsically, by definition of its balanced chemical reaction, produce wastes or by-products. Furthermore, atom economy focuses on the ideal situation of a chemical reaction that works exactly as written with 100% yield. As any undergraduate chemistry major will tell you, however, product is lost during transfers, side reactions lead to loss of material, and percent yield is rarely 100%. Moreover, in practice, an excess of one or more reagents is often used to drive a reaction to completion, and the excess reagent may not be recovered or recoverable.

A more practical and realistic metric is *reaction mass efficiency* (RME), defined as the actual mass of isolated product divided by the actual mass of all reactants used.[48]

$$\text{Reaction mass efficiency (RME)} = \frac{\text{Mass of product}}{\text{Mass of all reagents used}} \times 100\%$$

RME provides a more realistic picture of a reaction's efficiency because it uses the actual mass of the reactants used in a reaction and accounts for factors like the use of an excess of one (or more) reagents and reaction yield. However, calculation

* Catalysts, as well as solvents, are not traditionally considered in the atom economy calculation.

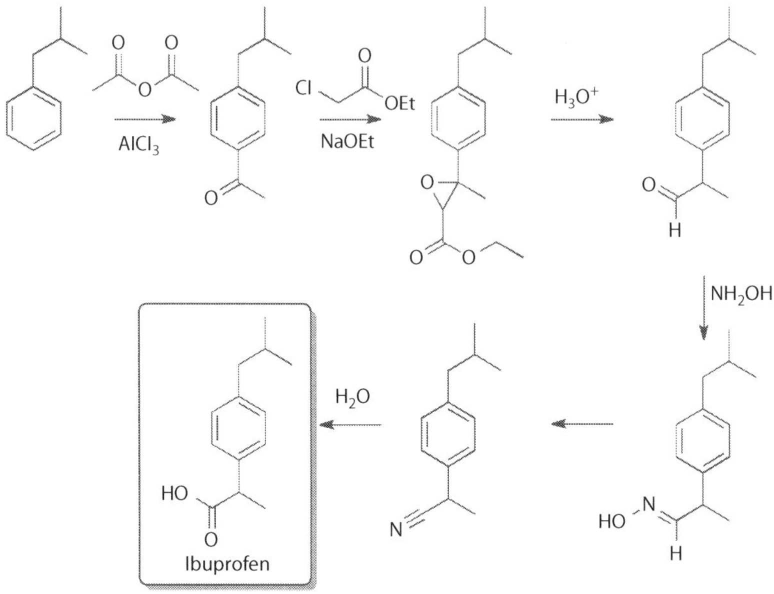

FIGURE 1.9 Boots synthesis of ibuprofen.

of RME requires that one have access to or knowledge of these laboratory values. Although RME is more realistic, atom economy is more easily calculated (especially for hypothetical reactions).

Where atom economy and RME focus on the effective use of chemical mass during a reaction, other metrics broaden the scope to take note of wastes generated during an overall process. Resources consumed during the workup of a reaction (i.e., the isolation and purification of a product), catalysts, and solvents are not normally considered in an atom economy or RME calculations. These materials are not reactants in a balanced chemical equation, but they are regardless consumed during a process (unless they are recovered and reused).

The E ("environmental") Factor accounts for this waste. E Factor, defined as the mass of waste produced per unit mass of product, is an applied metric that looks at the actual amount of waste generated by a process.[49,50]

$$\text{E Factor} = \frac{\text{Mass of total waste (kg)}}{\text{Mass of product (kg)}}$$

To calculate E Factor, one does not even need to know the chemical equation for the process of interest but can simply account for the masses of product formed and waste generated. In effect, how much product and how much "nonproduct" were produced?

Theoretically, a chemical reaction on paper may have an atom economy of 100%, yet still generate high amounts of waste if, for example, excess reagents are necessary, solvents are used, yields are low, or the product requires isolation or purification procedures (which are not included as part of a balanced chemical reaction).

Ideally, a process would have an E Factor of 0 (indicating no waste produced). In practice, E Factors can be as high as 100, although it can be difficult to accurately measure E Factor due to ambiguities in how one defines the waste from a procedure.

Determining how much waste is generated by a process first requires one to define what that "process" is precisely.[48] Clearly, waste generated on-site at the factory or plant counts as a result of the process. However, most chemical processes require energy (i.e., electricity). The generation of that energy produced waste at some remote location (i.e., the power plant). Should those wastes be included? They wouldn't have been generated had the process not been in operation. Because of ambiguities and variations in how different groups assess the wastes generated by a process, it can be difficult to directly compare E Factors from different sources without first investigating how the values were calculated.

A related metric, *process mass intensity*, is defined as the mass of all materials used in a procedure per mass of product.

$$\text{Process mass intensity (PMI)} = \frac{\text{Mass of all materials used (kg)}}{\text{Mass of product (kg)}}$$

PMI is related to the E Factor by the equation

$$\text{PMI} = \text{E Factor} + 1$$

Because they are related, a numerical E Factor essentially contains the same quantitative information as PMI; however, PMI is more commonly used in the pharmaceutical industry because it philosophically focuses on maximizing the efficient use of all resources put into a process.[51]

Although the "green" metrics described above are all useful for quantifying the waste generation and mass use of a chemical reaction or process, they each carry limitations.[48] The ambiguities in E Factor calculation were described above. Atom economy is simple to calculate but ignores any of the reality of a chemical reaction (e.g., percent yield). Process Mass Intensity is perhaps more realistic but requires one to have access to all of the carefully accounted inputs and outputs from a chemical process. Moreover, each metric may not necessarily inform the other. An evaluation of 28 different chemistries at GlaxoSmithKline identified that calculated values for metrics such as atom economy and reaction mass efficiency do not necessarily correlate.[52] Finally, the "green" metrics presented above all take the very harsh approach of considering everything as either "product" or "waste" (and further considering all wastes as equally bad). None of them evaluate the merits of a process based on its use of more or less dangerous substances or if a generated waste is better or worse than the alternative.

One example where metrics can be misleading is the case of epichlorohydrin. Epichlorohydrin, used chiefly for producing epoxy resins and also in the paper and pharmaceutical industries,[53] has been traditionally generated in a three-step procedure starting from propene (Figure 1.10).[54] An alternative pioneered at Dow Chemical Company instead begins with glycerine and employs a carboxylic acid

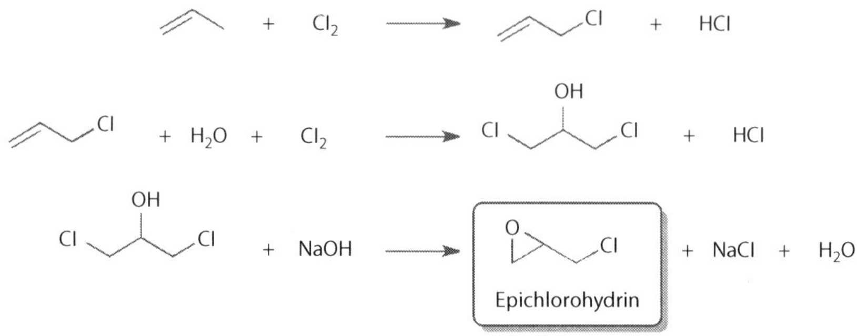

FIGURE 1.10 Synthesis of epichlorohydrin from propene. (Adapted from Bell, B.M. et al., *Clean* 36(8), 657–661, 2008.)

catalyst to generate the key, dichloropropanol intermediate (Figure 1.11).[54,55] (A similar process has also been commercialized by Solvay.[56]) Theoretically, epichlorohydrin production from propene or from glycerol have atom economies of 38% and 44%, respectively, suggesting little difference between them. However, where propene is a petrochemical from a nonrenewable feedstock, glycerine is a by-product of biodiesel production[57] (i.e., waste from biodiesel production becomes a useful material here). Additionally, the GTE ("glycerine to epichlorohydrin") route avoids the use of volatile chlorine gas. The "chlorine economy" of the GTE synthesis is also better; four chlorine atoms are required in the propene route whereas the GTE only requires two. Clearly, metrics can be misleading.

Metrics like atom economy and PMI allow one to easily and quickly evaluate the potential sustainability implications of a chemical process. However, they do not always do an adequate job in presenting the real costs, risks, and limitations of a procedure. Just as sustainability acknowledges the holistic interconnectedness of the Triple Bottom Line (people, planet, and profit) as crucial to meeting today and tomorrow's needs, so too does the field of life cycle assessment (LCA) recognize that it is folly to reduce a sustainable assessment to merely a single calculated metric.

LCA is a methodology for evaluating all parts of a process (chemical or otherwise) as an interconnected system. LCA considers a process in its own context,

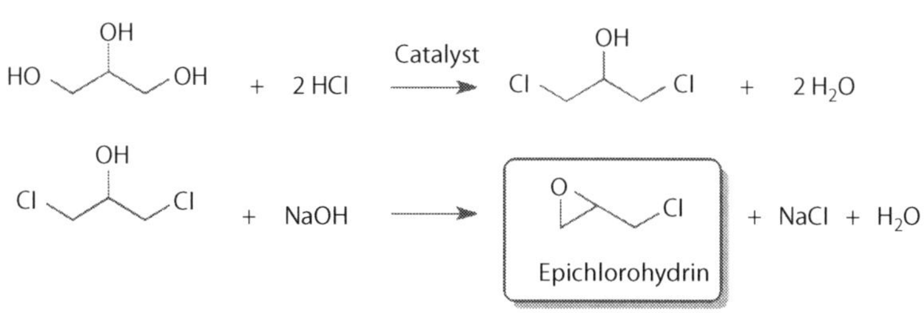

FIGURE 1.11 Synthesis of epichlorohydrin from glycerin. In practice, the first reaction produces a mixture of both 1,3-dichloro-2-propanol (shown) and 2,3-dichloro-1-propanol (not shown for clarity), both of which react in the second step to produce epichlorohydrin. (Adapted from Bell, B.M. et al., *Clean* 36(8), 657–661, 2008.)

evaluating all of the inputs (e.g., raw materials and energy sources) to a process, the environmental and economic costs at each step, and the ultimate fate of all generated substances (including not only pollution and wastes but also the product itself).[22] This comprehensive evaluation has been called a "cradle to grave" assessment.[58] Oftentimes, thorough LCA can reveal that a process initially appearing to be "green" on a small scale is actually less sustainable in practice. For example, a catalyst that allows the use of some renewable feedstock is no real advancement if the synthesis of that catalyst consumes more energy and fossil resources than the technology it was designed to replace.

Far too often, chemists report the development of some new, "green" technology that would not actually translate as a practical, sustainable solution beyond the research laboratory. This is not limited to scientists. In our increasingly environmentally concerned world, there are numerous examples of so-called "greenwashing" or overestimating the sustainable effects of an item.[59] Chemists wishing to make an impact in the realm of sustainability must be cognizant of the factors that are considered during a LCA while they are still at the design phase of their new technology. We cannot leave "sustainability" to be addressed only at the implantation phase of a technology.

1.6 GOALS, CHALLENGES, AND IMPACTS OF CATALYSTS

Traditional catalysts are being updated to respond to the demands of the Triple Bottom Line, and advances in catalytic technologies now directly confront challenges raised by sustainability (such as pollution remediation and the development of alternative energy sources). The chapters that follow will explore the principal fields of catalysis: metal-based catalysis, metal-free organocatalysis, and biocatalysis (enzymes). We also address the synthetic catalysts inspired by nature and enzymes, the use of simple, abundant, and affordable clays to catalyze organic reactions, and the development of catalysts necessary for water oxidation, artificial photosynthesis, and the future of our global energy economy. As we've seen in this chapter, however, the chemist must always be wary of overestimating and misinterpreting the sustainable implications of laboratory-scale discoveries. Therefore, the final chapter of *Catalysis for Sustainability* provides an overview of LCA and applies it to a number of proposed or in-practice catalytic technologies.

Catalysts have proven their utility on a global scale, being ubiquitous to our goods, services, and general way of life. Moreover, catalysts have an inherent potential to address the needs of the environment, the economy, and all of society. In the years to come, it will be exciting to see how catalysts (continue to) have a worldwide effect on our journey toward sustainable development.

REFERENCES

1. World Commission on Environment and Development. *Our Common Future*. Oxford University Press: Oxford; New York, 1987.
2. Elkington, J. *Cannibals with Forks: The Triple Bottom Line of 21st Century Business*. New Society Publishers: Gabriola Island, BC; Stony Creek, CT, 1998.

3. Willard, B. *The Sustainability Advantage: Seven Business Case Benefits of a Triple Bottom Line*. New Society Publishers: Gabriola, BC, 2002.
4. Anastas, P.T., and Warner, J.C. *Green Chemistry: Theory and Practice*. Oxford University Press: Oxford, UK, 1998.
5. Speight, J.G. *The Chemistry and Technology of Petroleum*, 5th ed. CRC Press, Taylor & Francis Group: Boca Raton, FL, 2014.
6. U.S. Energy Information Administration. Available at http://www.eia.gov/ (accessed November 15, 2014).
7. Petrochemicals. Available at http://www.afpm.org/petrochemicals/ (accessed November 15, 2014).
8. Hooper, D.U., Chapin, F.S., Ewel, J.J., Hector, A., Inchausti, P., Lavorel, S., Lawton, J.H. et al. *Ecological Monographs* 2005, 75 (1), 3–35.
9. Rands, M.R.W., Adams, W.M., Bennun, L., Butchart, S.H.M., Clements, A., Coomes, D., Entwistle, A. et al. *Science* 2010, 329 (5997), 1298–1303.
10. Breslow, R. *The Scientist* 1993, 7 (6), 11.
11. Sanderson, K. What are you afraid of? *Chemistry World* 2013. Available at http://www.rsc.org/chemistryworld/2013/10/chemophobia.
12. Kovács, L., Csupor, D., Lente, G., and Gunda, T. *100 Chemical Myths*. Springer: New York, 2014.
13. Carson, R. *Silent Spring*, 1st ed. Fawcett Crest: New York, 1964.
14. Bomgardner, M.M. *Chemical Engineering News* 2012, 90 (11), 30–32.
15. McDonough, W., and Braungart, M. *Cradle to Cradle: Remaking the Way We Make Things*, 1st ed. North Point Press: New York, 2002.
16. Centi, G., and Perathoner, S. From green to sustainable industrial chemistry. In: *Sustainable Industrial Chemistry*, Edited by Cavani, F., Centi, G., Perathoner, S., and Trifiró, F. Wiley-VCH: Weinheim, Germany, 2009.
17. Green Chemistry. Available at http://www2.epa.gov/green-chemistry (accessed November 14, 2014).
18. Ritter, S.K. *Chemical Engineering News* 2002, 80 (26), 26–30.
19. Ritter, S. *Chemical Engineering News* 2011, 89 (26), 11.
20. IUPAC. *Compendium of Chemical Terminology*, 2nd ed. (The "Gold Book"). Compiled by A. D. McNaught and A. Wilkinson. Blackwell Scientific Publications, Oxford, 1997. XML on-line corrected version: http://goldbook.iupac.org (2006) created by Nic, M., Jirat, J., and Kosata, B., updates compiled by A. Jenkins. doi:10.1351/goldbook.
21. Simmons, M.S. The role of catalysts in environmentally benign synthesis of chemicals. In: *Green Chemistry: Designing Chemistry for the Environment*, Edited by Anastas, P.T., and Williamson, T.C. American Chemical Society: Washington, DC, 1996.
22. Lancaster, M. *Green Chemistry: An Introductory Text*, 2nd ed. Royal Society of Chemistry: Cambridge, 2002.
23. Haber, F., and Rossignol, R.L. *Journal of Industrial & Engineering Chemistry* 1913, 5 (4), 328–331.
24. Erisman, J.W., Sutton, M.A., Galloway, J., Klimont, Z., and Winiwarter, W. *Nature Geoscience* 2008, 1 (10), 636–639.
25. Doyle, M.W., Stanley, E.H., Havlick, D.G., Kaiser, M.J., Steinbach, G., Graf, W.L., Galloway, G.E., and Riggsbee, J.A. *Science* 2008, 319 (5861), 286–287.
26. U.S. Geological Survey, Mineral Commodity Summaries 2014, p. 196. Available at http://minerals.usgs.gov/minerals/pubs/mcs/2014/mcs2014.pdf.
27. Smil, V. *Enriching the Earth: Fritz Haber, Carl Bosch, and the Transformation of World Food Production*. MIT Press: Cambridge, MA, 2001.
28. Smil, V. *Nature* 1999, 400 (6743), 415.
29. Gates, B.C. *Catalytic Chemistry*. Wiley: New York, 1992.

30. Beller, M., Renken, A., and Santen, R.A.V. *Catalysis: From Principles to Applications*. Wiley-VCH Verlag GmbH & Co. KGaA: Weinheim, Germany, 2012.
31. Anslyn, E.V., and Dougherty, D.A. *Modern Physical Organic Chemistry*. University Science: Sausalito, CA, 2006.
32. Hamori, E., and Muldrey, J.E. *Journal of Chemical Education* 1984, 61 (8), 710.
33. Centi, G., and Perathoner, S. Methods and tools of sustainable industrial chemistry: Catalysis. In: *Sustainable Industrial Processes*, Edited by Cavani, F., Centi, G., Perathoner, S., and Trifiró, F. Wiley-VCH: Weinheim, Germany, 2009.
34. Hagen, J. *Industrial Catalysis: A Practical Approach*, 2nd ed. Wiley-VCH: Weinheim, Germany, 2006.
35. Sato, K., Aoki, M., and Noyori, R. *Science* 1998, 281 (5383), 1646–1647.
36. Van de Vyver, S., and Roman-Leshkov, Y. *Catalysis Science & Technology* 2013, 3 (6), 1465–1479.
37. Cavani, F., and Teles, J.H. *ChemSusChem* 2009, 2 (6), 508–534.
38. Kumar, A.S., and Prasad, P.S.S. Cracking and oxidative dehydrogenation of ethane to ethylene. In: *Industrial Catalysis and Separations: Innovations for Process Intensification*, Edited by Raghavan, K.V., and Reddy, B.M. Apple Academic Press: Toronto, 2014.
39. Carey, F.A., and Sundberg, R.J. *Advanced Organic Chemistry: Part A: Structure and Mechanisms*, 5th ed. Springer: New York, 2007.
40. Ma, S.K., Gruber, J., Davis, C., Newman, L., Gray, D., Wang, A., Grate, J., Huisman, G.W., and Sheldon, R.A. *Green Chemistry* 2010, 12 (1), 81–86.
41. Jimenez-Gonzalez, C., Constable, D.J.C., and Ponder, C.S. *Chemical Society Reviews* 2012, 41 (4), 1485–1498.
42. Trost, B.M. *Science* 1991, 254 (5037), 1471–1477.
43. Trost, B.M. *Angewandte Chemie International Edition* 1995, 34 (3), 259–281.
44. Sheldon, R.A. *Chemical Society Reviews* 2012, 41 (4), 1437–1451.
45. Elango, V., Murphy, M.A., Smith, B.L., Davenport, K.G., Mott, G.N., Zey, E.G., and Moss, G.L. Method for producing ibuprofen. U.S. Patent 4,981,995, January 1, 1991.
46. Cann, M.C., and Connelly, M.E. *Real-World Cases in Green Chemistry*. American Chemical Society: Washington, DC, 2000.
47. Doble, M., and Kruthiventi, A.K. *Green Chemistry and Engineering*. Elsevier Academic Press: Burlington, MA, 2007.
48. Constable, D.J.C., Curzons, A.D., and Cunningham, V.L. *Green Chemistry* 2002, 4 (6), 521–527.
49. Sheldon, R.A. *Green Chemistry* 2007, 9 (12), 1273–1283.
50. Sheldon, R.A. *Chemical Communications* 2008 (29), 3352–3365.
51. Jimenez-Gonzalez, C., Ponder, C.S., Broxterman, Q.B., and Manley, J.B. *Organic Process Research & Development* 2011, 15 (4), 912–917.
52. Curzons, A.D., Constable, D.J.C., Mortimer, D.N., and Cunningham, V.L. *Green Chemistry* 2001, 3 (1), 1–6.
53. Smiley, R.A., and Jackson, H.L. *Chemistry and the Chemical Industry: A Practical Guide for Non-Chemists*. CRC Press: Boca Raton, FL, 2002.
54. Bell, B.M., Briggs, J.R., Campbell, R.M., Chambers, S.M., Gaarenstroom, P.D., Hippler, J.G., Hook, B.D. et al. *Clean* 2008, 36 (8), 657–661.
55. Hogue, C. *Chemical & Engineering News*, 2012, 90 (5), 52.
56. McCoy, M. *Chemical Engineering News* 2012, 90 (10), 15.
57. Leung, D.Y.C., Wu, X., and Leung, M.K.H. *Applied Energy* 2010, 87 (4), 1083–1095.
58. Horne, R., Grant, T., and Verghese, K. *Life Cycle Assessment: Principles, Practice, and Prospects*. CSIRO: Collingwood, Australia, 2009.
59. Dahl, R. *Environmental Health Perspectives* 2010, 118 (6), A246–A252.

2 Transition Metal Catalysis for Organic Synthesis

Jeremy R. Andreatta and Meghna Dilip

CONTENTS

The development of new catalysts and catalytic processes are vitally important to the "green" synthesis of commodity chemicals. A significant piece in the achievement of the goals of green chemistry, the use of catalysts, is enshrined in principle no. 9 of the Twelve Principles of Green Chemistry:[1]

> Catalytic reagents (as selective as possible) are superior to stoichiometric reagents.

With the initial publication of the tenets of green chemistry in the late 1990s came a challenge to develop new pathways of chemical synthesis that focused on preventing the formation of waste and maximizing atom economy (incorporation of all atoms in the starting product into the end product). Before the development of green chemistry, industries often looked to achieve a high percentage yield with very little regard for the amount of by-products that could accumulate along the synthetic pathway.[2] With the institution of green chemistry, the focus changed from maximizing the mass of product obtained to maximizing the number of atoms from the starting material that are incorporated into the desired product. The ideal synthetic process will have both a high yield and high atom economy. One way to decrease the number of atoms wasted is to perform syntheses/transformations using a catalyst.

A highly selective, nonstoichiometric, reusable catalyst reduces the formation of by-products and does not contribute itself to waste production, thus significantly improving atom economy. It should be noted that a high atom economy in itself does not imply a green process because atom economy does not take into consideration the toxicity of reagents or by-products, energy requirements, and type or amount of solvents used, among other things. Another major benefit of using a catalyst is the gains achieved in energy efficiency. Because the use of a catalyst lowers the energy of activation of a chemical process, transformations can be carried out at lower temperatures with a lower energy outlay, thus satisfying yet another principle of green chemistry, that of increased energy efficiency (principle no. 6).[1]

> Energy requirements of chemical processes should be recognized for their environmental and economic impacts and should be minimized. If possible, synthetic methods should be conducted at ambient temperature and pressure.

- Although there are intrinsic green benefits with the use of catalysts (increased atom economy, lower temperatures of operation), to truly achieve the goals of green chemistry, the catalyst should be highly selective and only catalyze the formation of the desired product and minimize the formation of any by-products or harmful waste.
- The catalyst should have a high turnover frequency (ToF) and turnover number (ToN). It should function at low concentrations and ideally be recyclable or have a long lifetime to minimize long-term costs and waste production.
- The catalyst should be readily compatible/active in green solvents or adaptable to solventless reaction conditions.
- The catalyst should be easily isolated from the reaction medium/vessel to allow for lower overall process energy inputs and purification costs.
- Any metal (i.e., transition metal) catalysts should be derived from easily mined, earth-abundant sources that are minimally toxic at any stage of the process including the refining process and when in its final catalyst form.

Organic reactions have been one of the most important applications of transition metal catalysts and have served as a major driving force in the area of transition metal inorganic and organometallic chemistry (the chemistry of metal–carbon bonds).[3–5] Transition metal complexes have played, and will continue to play, a major role in the development of greener processes for the production of the compounds and products we use throughout our everyday lives.[6]

Transition metals are especially attractive as catalysts due to their partially filled d-orbitals, which allow them to bind a variety of substrates and perform complex electronic transformations. Additionally, where carbon and many other elements can hold a maximum of 8 electrons (or 12 in the case of hypervalent later main group elements), transition metals can accommodate up to 18 valence electrons in orbitals that are relatively high in energy and allow for electrons to be readily transferred in and out (off and onto the metal).[7] This allows for reactants to interact with a transition metal, undergo a transformation, and then be released so that the complex can reenter the catalytic cycle.

What follows will focus on a selection of transition metal–catalyzed chemical reactions that are currently used in industrial scale processes. Specifically, it will focus on processes that are catalyzed by well-defined single-site homogeneous catalysts and the general mechanisms by which they operate. Each discussion will aim to inform of the fundamental steps involved in the catalytic cycle along with the different variations that allow for a varied substrate scope and the production of a wide variety of products. Following the description of the current methods, we will investigate some developments for more sustainable/green versions for each of the processes. One area of particular interest in most academic and industrial laboratories is the development of catalysts that use earth-abundant metals such as iron or manganese.

The metal is not the only piece of the catalyst puzzle. The ligand (i.e., the organic group that surrounds/is attached to the active transition metal center) affects the rate and specificity (stereochemistry or regiochemistry) of all catalytic processes. The seemingly infinite combinations/permutations of the ligands involved in a catalytic could fill several book chapters by themselves. The development of modularly synthesized or simple ligands or a combinatorial chemical approach to catalysis is becoming the status quo for industrial catalysis development and is beyond the scope of this chapter. The ligand(s) used in each of the following applications play a major role and are a significant source of capital investment in the development of new catalysts, many times much more expensive than the metal precursor itself. For that reason, we will focus on the general catalytic cycles/processes and allude to the ligands occasionally as necessary.

2.1 FORMATION OF CARBON–CARBON BONDS

The formation of carbon–carbon bonds is essential for the effective synthesis of many of the chemicals we encounter in everyday life. Many of the processes to produce both pharmaceuticals and bulk chemicals require the construction of at least one, if not several, new carbon–carbon bonds. Early reports of C–C bond-forming reactions involved harsh reaction conditions (such as high temperatures) and the use of stoichiometric reagents such as Grignards (e.g., CH_3MgBr), alkyl lithium, alkyl zinc, or Wittig reagents ($(Ph)_3PCR_2$).[8] In contrast to these stoichiometric reagents, catalyzed processes typically perform similar couplings while producing less waste and minimizing energy input. Many transition metal–catalyzed processes use relatively non–earth abundant and expensive metals from the platinum group metals (PGMs, Rh, Pd, Pt, Rh, etc.). However, the high activity of these reagents permits low catalyst loading (typically 0.1–0.5 mol%) and is highly selective for the synthesis of a variety of products, which offsets the high cost of the metal source. Excellent reviews of the specific ligands and complexes that are used in cross-coupling already exist and will not be further discussed herein.[9–12]

2.1.1 Cross-Coupling

Arguably one of the most utilized C–C bond forming reactions to date is the cross-coupling of two carbon sources using palladium catalysts. Pd catalyzed cross-coupling finds utility in both bulk and small-scale chemical production.[13] Palladium has emerged as the metal of choice in this and other transition metal–catalyzed processes due to its ability to activate a wide variety of substrates at high ToN and usually at lower temperatures than its more earth abundant and cheaper non–precious metal counterparts.

At its heart, carbon–carbon cross-coupling reactions involve the formation of a bond between an electrophilic carbon source (R–X) and a nucleophilic carbon source (R–FG; Figure 2.1). The importance of C–C bond forming was recognized in 2010 when Heck, Negishi, and Suzuki were awarded the Nobel Prize in Chemistry for their work in this area.[14]

A number of variations of carbon–heteroatom (noncarbon) cross-coupling reactions are known, including Hartwig–Buchwald coupling (C–N bond formation), carbonylative coupling to form ketones (CO), and others.[15–18] Many of these coupling reactions have only recently been reported and have yet to be developed on an industrial scale. Undoubtedly, as more is understood about these processes, they will find use in a variety of industrial-scale processes. These reactions are extremely versatile and can be used to form not just carbon–carbon but also C–heteroatom bonds in a wide variety of substrates. For example, cross-coupling using CO as the nucleophilic R source has been used in the synthesis of ibuprofen, saving both time (by reducing the total number of synthetic steps) and reducing the overall waste production.[19]

For any transition metal–catalyzed process, it is important to understand the mechanism. It is only through this deeper understanding of how each component affects the reaction that a process can be made more efficient or green in nature (or both). It serves to streamline the discovery process by providing the information to make informed decisions of how to modify the process to be more environmentally benign. To that end, we will first discuss the general mechanism for palladium catalyzed cross-coupling.

In the first step of the catalytic cycle, the alkyl halide (R–X) oxidatively adds to a low coordination number Pd^0 center yielding the $Pd^{II}(R)(X)$ complex. The second major step of the catalytic cycle installs a second carbon-containing group, R′, onto the Pd center and produces one equivalent of waste FG–X. A variety of R′–FG complexes have been studied as possible reagents and as sources of nucleophilic carbon. The final and product-releasing step of the catalytic cycle is the reductive elimination of the product, which regenerates the low coordinate Pd(0) catalyst that can reenter the cycle (Scheme 2.1).

For the initial oxidative addition to take place, the palladium metal complex has to be sufficiently electron-rich enough to nucleophilically attack the alkyl halide (R–X) substrate. On the other hand, during the reductive elimination stage, the ligand should be able to leave easily to allow the formation of a C–C bond. The catalysts used for this process must trace a fine line between being able to activate the R–X bond (oxidative addition) while simultaneously being able to reductively eliminate to form the new C–C bond. The requirements for both the initial step in

$$\text{R–X} + \text{R'–FG} \xrightarrow{\text{cat.}} \text{R–R'}$$

$\overset{\delta+\ \delta-}{\text{R–X}}$
Electrophilic R
(X = Cl, Br, I, OTf)

$\overset{\delta-\ \delta+}{\text{R'–FG}}$
Nucleophilic R
(FG = $B(OR)_2$, ZnX, SnR_3, CH_nR_{2-n}, $CH_2(COOMe)_2$)

FIGURE 2.1 Carbon reagents for Pd-catalyzed cross-coupling reactions.

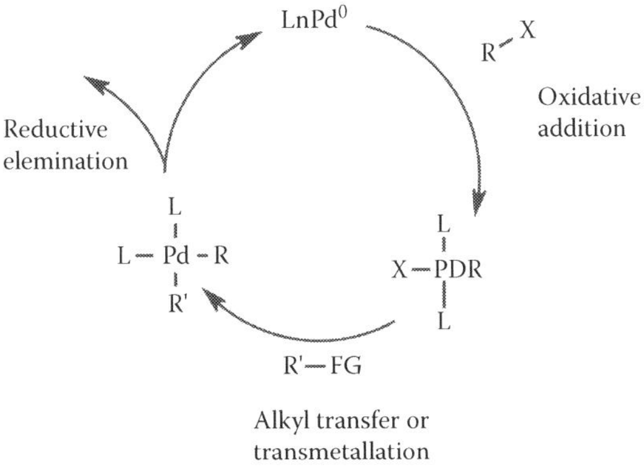

SCHEME 2.1 General mechanism for Pd-catalyzed C–C bond formation. FG = functional group, R = alkyl or aryl, X = halide.

the catalytic cycle and the product-releasing final step will be discussed further in the following section.

Most Pd catalyzed cross-coupling reactions use Pd complexes, such as palladium(II) acetate combined *in situ* with bulky, strong electron-donating ligands such as phosphines (PR_3) or in some instances N-heterocyclic carbenes (NHCs) to generate the necessary electron-rich low-coordinate Pd^0 active species such as $Pd(PR_3)_2$ or Pd(NHC) (Figure 2.2).[20] The use of strongly electron-donating ligands is necessary to not only stabilize the initial Pd^0 complex but also to facilitate the oxidative addition of R–X to the Pd center. The use of bulky Lewis basic phosphine ligands is also important for the final product–releasing reductive elimination step *vide infra*.[21]

After the formation of the Pd^{II} species, $L_2Pd(X)(R)$, we encounter a potential off cycle (i.e., undesired) reaction where a substrate can undergo intramolecular β-hydride activation that can lead to undesired by-products (Scheme 2.2) and a deactivation of the Pd catalyst. This is the major reason why many cross-coupling reactions are limited to the use of aryl or vinyl as the electrophilic carbon source; however, recent advances have allowed for the use of some alkylhalides.[22–26] For example, Fu and coworkers[27] have developed nickel catalysts that promote both the Suzuki and Negishi cross-coupling of haloalkane substrates.

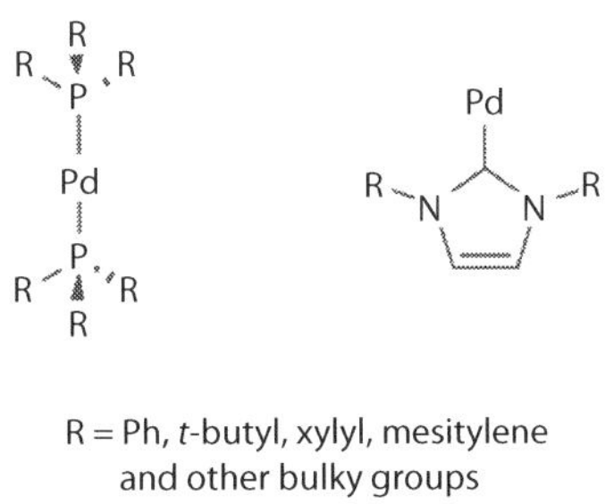

FIGURE 2.2 Active forms of Pd catalysts with bulky electron-donating ligands for cross-coupling.

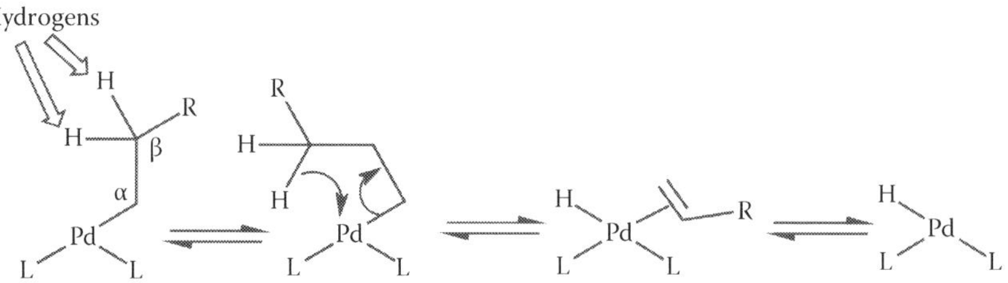

SCHEME 2.2 β-Hydride elimination during Pd catalyzed cross-coupling.

One advantage to the use of phosphine ligands over NHCs or other ligand types is the wide varieties that are commercially available with varying steric and electronic properties. For example, kinetic studies have found that catalysts bearing more sterically hindered phosphines (i.e., $P(^tBu)_3$) are actually faster for the oxidative addition step of the catalytic cycle.[28,29] This is attributed to the ability of the bulkier ligand being able to stabilize the low d-electron count, coordinately unsaturated PdL_2 species in solution versus less sterically demanding ligands such as PMe_3. A seemingly countless number of ligands have been developed for cross-coupling that allow for highly regioselective and stereoselective couplings and also to develop catalysts that are tolerant to a variety of X groups on the electrophilic carbon source.[30–32] As will be discussed later in this chapter, transition metal catalysts have since been developed that can form carbon–carbon bonds using substrates that lack any functionalization (i.e., a halogen substituent X).

The next step in the cross-coupling reaction, alkyl transfer/transmetallation, is where the major distinction among the named reactions can be made (Table 2.1). In

TABLE 2.1
Named Pd Catalyzed Cross-Coupling Reactions and Comparison of Nucleophilic Carbon Sources

Named Reaction	R′-FG	Pros and Cons of Various Cross-Coupling Pathways
Suzuki–Miyaura	$R'\text{-}B(OH)_2$	con—Boronic acid/esters must be synthesized separately
Negishi	R′-ZnX	con—Zinc reagents are air and moisture sensitive
Sonogashira	Cu-CCR′	pro—Catalytic in both Pd and Cu
Heck–Mizoroki	$H_2C = CHR'$	con—Needs base present to complete the catalytic cycle and regenerate Pd^0 con—R needs to be electron withdrawing group to ensure regiocontrol
Stille	R_3SnR' (R = alkyl)	con—Alkyl tin reagents can be highly toxic
Tsuji–Trost	$CH_2(COOMe)_2$	con—Limited to allylic acetates con—Needs base to activate the R′-FG
Hiyama–Tamo	$R'Si(E)_3$ (E = F_nR_{2-n} or OR)	con—Requires addition of F^- in most cases to activate the silane

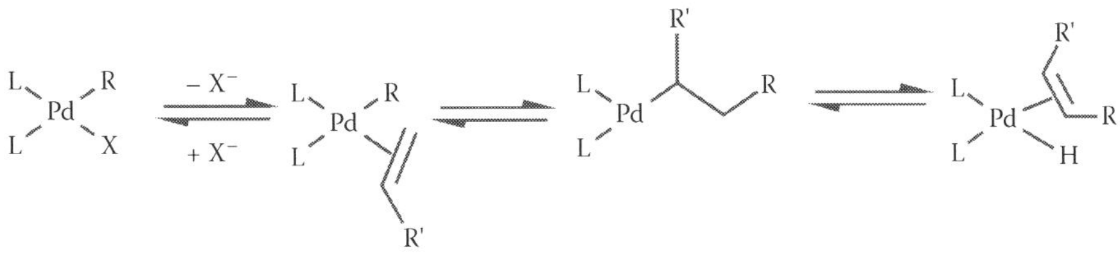

SCHEME 2.3 Heck–Mizoroki cross-coupling alkyl transfer pathway. The alkene is drawn in the *cis*- conformation for ease of viewing purposes only.

this step of the catalytic cycle, the Pd–X bond is broken and replaced with a second alkyl or aryl group (R′).

Although the source of nucleophilic R′ varies, the mechanism for all but the Heck–Mizoroki coupling remains the same. The nucleophilic substrate undergoes a transmetallation (the organometallic version of salt metathesis) in which a new Pd-R′ forms a bond along with the production of one equivalent of a halogenated waste product. This step is usually irreversible due to the thermodynamic or kinetic stability of the newly formed C–C bond.

The Heck–Mizoroki (aka Heck reaction) differs in that it does not require a transmetallation step. The olefin used is sufficiently nucleophilic in nature to attack the Pd center and displace either the X or L ligand. It then inserts into the Pd-R bond to form a new Pd-σ-alkyl species. This step is followed by β-hydride elimination to form the $Pd^{II}(H)(\eta^2$-olefin) complex before the product release step (Scheme 2.3). After the release of the product, the remaining Pd–hydride complex (if necessary) binds X^- and reductively eliminates HX. This is also different from the other cross-coupling reactions in that reductive elimination is not the product-forming step of the reaction, but serves only to regenerate the catalyst.

Regardless of substrate, the final step of each catalytic cycle involves reductive elimination of either an organic product or HX from Pd^{II} to regenerate the initial catalytically active Pd^0 complex. Again, as with the oxidative addition step, this is facilitated by the ligands that surround the Pd center. Whereas the electron-rich center facilitated the oxidative addition in the first step of the catalytic cycle, here the steric bulk of the ligand provides the driving force for the release of the product.* One might be curious how the PdL_2 fragment that was electron-rich enough to facilitate oxidative addition earlier in the reaction is now able to undergo reductive elimination, which is the opposite of oxidative addition. The answer lies in the mechanism of the reductive elimination step, which has been studied in depth for many Pd^{II} complexes (Scheme 2.4). The findings demonstrate the beauty and complexity that lies behind the simplified catalytic cycle in Scheme 2.1. In general, Pd complexes with multiple monodentate ligands, L, will actually dissociate one ligand (L) before the reductive elimination step. The reductive elimination from the three-coordinate LPd(R)(R′) requires significantly less energy that the same process performed from the 4 coordinate $L_2Pd(R)(R')$. This is, at least in part, due to the reduced electron density in the LPd versus the L_2Pd (i.e., fewer electron donors means less electron-rich Pd).

* A *cis* configuration of the R and R′ (or H and X for Heck) ligands is required for this step.

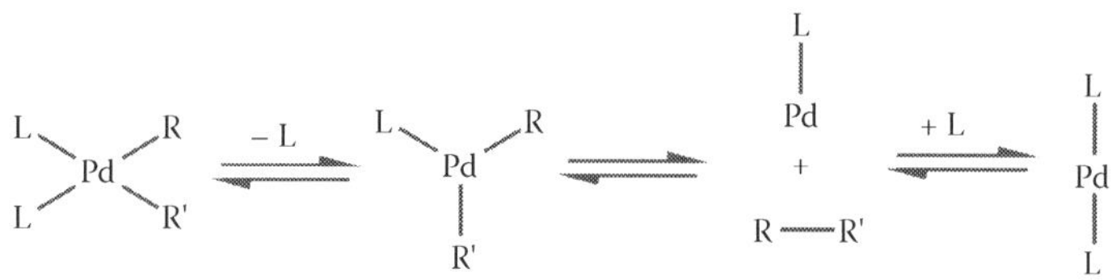

SCHEME 2.4 Mechanism of reductive elimination from $L_2Pd(R)(R')$.

Additionally, a molecular orbital explanation for the buildup of antibonding interactions in L_2Pd as the R–R bond forms increases the activation energy of reductive elimination, which is avoided to a great extent in the three-coordinate case. One must be careful not to broadly apply this pathway to all reductive elimination processes (and all Pd-based systems). It is highly dependent on the ligand(s), the identity of the transition metal, and the oxidation state of the transition metal.[33] The in-depth studies of the reductive elimination step of the catalytic cycle through both wet chemical and computational methods has led to a deeper understanding of the factors that affect the overall performance of the palladium complexes. From these studies, we know that the addition of excess phosphine or NHC ligands to the reaction will actually diminish productivity and lead to the generation of more waste in the form of unused ligand.

The applications of Pd-catalyzed cross-coupling are widespread throughout the chemical industry.[12,34–36] A few examples include the synthesis of Taxol® (via Heck), agricultural chemicals (e.g., herbicides), natural product synthesis like the natural marine antiviral hennoxazole A (Negishi), and the antifungal Terbinafine (Suzuki–Miyaura; Figure 2.3).[37–39] Since the late 1990s, the number of publications and patents for the different types of cross-coupling reactions has rapidly increased (Graph 2.1, black line).[29]

Although the power of C–C coupling is well respected, the study of cross-coupling reactions that meet the needs of green chemistry and sustainability is still relatively new.[40] Although the number of reports is somewhat limited, an increased interest in

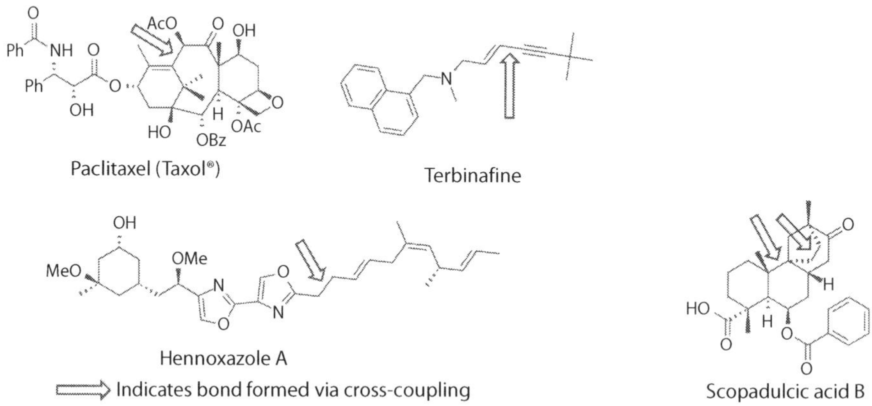

FIGURE 2.3 Selected compounds synthesized using cross-coupling.

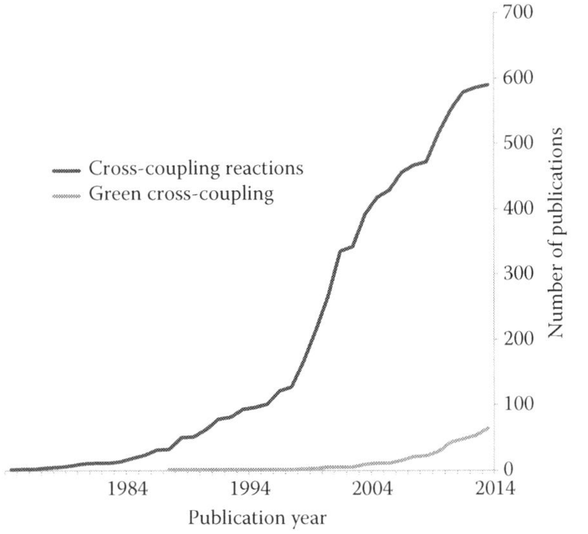

GRAPH 2.1 Plot of number of publications for "palladium catalyzed cross-coupling" per year. This graph is derived from a subject search of "palladium catalyzed cross-coupling" (black) and "green cross-coupling" (gray) using the Scifinder database.

green cross-coupling began in the late 1990s (Graph 2.1, gray line). Cross-coupling is a catalyzed process and therefore is inherently green, but there is still much room for improvement in the areas of achieving reactions at lower temperatures (current reactions are carried out at 80°C to 110°C) and shorter reaction times are required in many cases. Other areas of improvement include the need to develop catalysts that do not require activated substrates in the form of aryl halides (as the electrophilic R source) and the variety of nucleophilic carbon sources (R′-FG) that cumulatively create significant amounts of salt waste. Additionally, the use of halogenated organic solvents, the need for substituted substrates (alkyl/aryl halides, etc.), and use of platinum group metals that can be both expensive and difficult to recover/separate from the reaction medium during the product isolation procedure(s) are all areas needing attention.

Traditionally, cross-coupling reactions are carried out in polar organic solvents such as toluene, acetonitrile, methanol, dimethylformamide, or tetrahydrofuran.[41,42] According to the GSK solvent selection guide, originally developed for easy solvent selection in the pharmaceutical industry, tetrahydrofuran, dimethylformamide, and acetonitrile are solvents that have "major issues" and should be avoided.[43] Issues with these solvents include high volatility, high flammability, and human and environmental toxicity issues. Greener options are alcohols such as methanol and ethanol, which are highly stable solvents and have a low environmental impact. On the downside, alcohols are not easily recycled because they require large amounts of energy for distillation and purification. One way to overcome this issue is to use alcohol–water mixtures rather than pure alcohols themselves. Other options such as ethyl lactate, glycerol, and polyethylene glycol are nontoxic and are also preferred solvents. Ionic liquids, defined as organic salts that are liquid at room temperature, are a large class of solvents receiving much attention for their nonvolatility and thus limited release into the air.[44] However, the proper choice of ionic liquid determines the overall greenness of the solvent.[45,46] From a purely solvent selection perspective, a reaction is best performed in water.[47] Water is nontoxic, nonflammable, cheap, and (for the time being) abundant. With several options to consider, it is important to view the role of solvent holistically in the reaction. For instance, although water is the best solvent from a green perspective, cross-coupling in pure water without any additional solvents, modified ligand architectures, or surfactants/additives has been elusive (*vide infra*). Water with even small amounts of organic residue is often considered as "organic waste," and such waste will need expensive separation rather than simple incineration for disposal after use. Yet another option is to use no solvent at all (i.e., solventless reactions). However, the scale-up of these reactions can be extremely unsafe as they present the risk of formation of "hot spots" in the reaction vessel and runaway reactions.[48,49]

The temperature at which a reaction is carried out is another consideration. Consider, for example, a recent report of cross-coupling in water using a palladacycle catalyst operating at an elevated temperature of 100°C (Scheme 2.5).[50] This does indeed perform cross-coupling in water; however, it requires a reaction temperature equivalent to or higher than 100°C. Not only does the higher reaction temperatures mean more energy and expense, but the elevated temperatures cause corrosion of the reaction vessel to be a genuine issue. Additionally, in this process, it was seen that the catalyst could not be recycled efficiently beyond four cycles. The catalyst itself needed modification to make it soluble in water, the synthesis of which needs a separate green evaluation (e.g., how many steps to synthesis, were protecting and deprotecting groups used). Thus, although purely from a solvent perspective, this reaction would be labeled "green," a deeper examination reveals issues. In yet another example, cross-coupling was demonstrated using the green solvent mixture of ethyl lactate and water (Scheme 2.6).[51] Advantages include the use of a relatively simple palladium catalyst with high yields of the desired product. However, large amounts of organic solvent were then used to recover the product. The use of supercritical

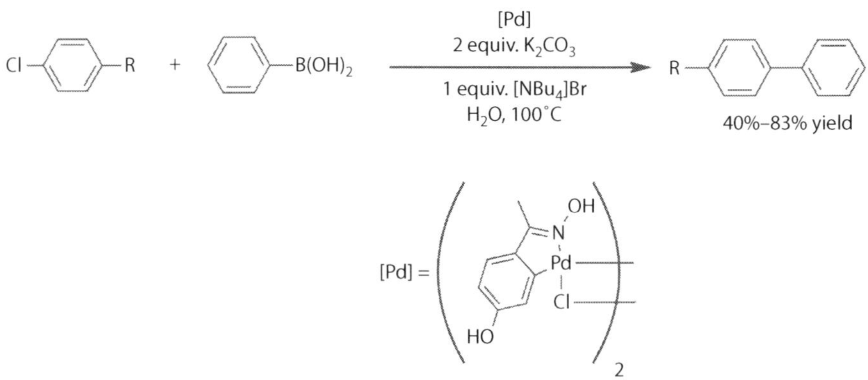

SCHEME 2.5 Suzuki–Miyaura synthesis of biaryls using palladocycle oxime catalyst in H_2O.

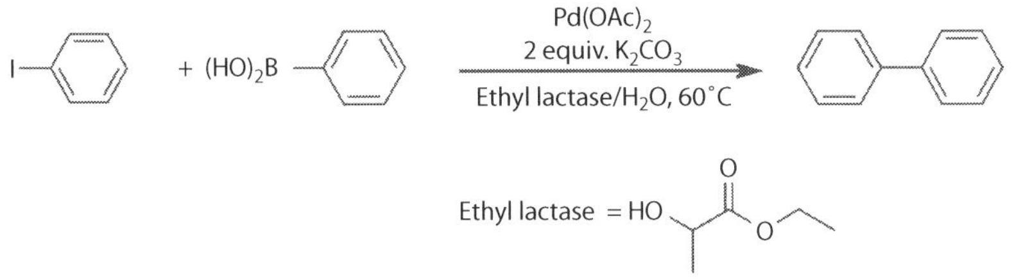

SCHEME 2.6 Suzuki–Miyaura reaction using a simple palladium catalyst in ethyl lactate/water solvent.

CO_2 has been examined to recover product and represents a way to avoid the use of organic solvents and can make the overall process greener.[52] It is thus hard to simply pinpoint the use of a green solvent and label the reaction green. While choosing a solvent, one should pay attention to other reaction parameters such as temperature, time of reaction, relative yields, and catalyst loading as well as the overall sustainability of ligand modifications needed before making an informed choice (in other words, follow a life cycle analysis approach).

Solvent selection also depends on the reaction conditions chosen. Alternate reaction conditions in lieu of thermal heating such as microwave heating and use of ultrasound can accelerate reactions and reduce reaction time, and also frequently allow for lower catalyst loading.[53–55] For effective microwave heating solvents with a high tan δ (where δ is the ratio of dielectric loss to dielectric constant of the solvent) is desired. Pure water has a medium value of 0.123 and, for most effective microwave heating, would require the addition of salts or other solvents. The exact mechanism/effect of microwaves on a reaction is still a matter of debate as is the overall safety and scalability of microwave technology to industrial scales. These issues will ultimately determine the adoption of microwave heating on a large scale.[56]

Modification of ligands in general to increase solvent-solubility is expensive but routes that include the use of renewable non–petroleum-derived ligands, such as sugars, are greener alternatives and need to be investigated further.[47,57] Another option is to add solubilizing ionic groups such as sulfates or ammonium groups to the ligand architecture (Figure 2.4).[58,59] At the very least, this is an expensive process and could potentially cause problems during the reaction due to the addition of cations/anions to the reaction medium.

An alternative to making water-soluble ligands/complexes is the addition of micelle-forming surfactants to facilitate the reaction in aqueous medium.[60,61] These operate much like soaps that form capsules (aka micelles) around oily/organic (non–water soluble) stains in clothing to be washed away in aqueous solution. Here, the surfactants form microreactors around the nonsoluble reagents to facilitate the reaction. A limited variety of commercially available surfactants have shown promise in the area of Pd catalyzed cross-coupling, which facilitates the reaction not only in aqueous media but also at room temperature, which is another advantage. It should be noted that the surfactant concentrations are very low (usually ~2–8 wt%). Much of the work in the area is focused on the development of new surfactants that will hopefully widen the applicability of these applications. Several examples of

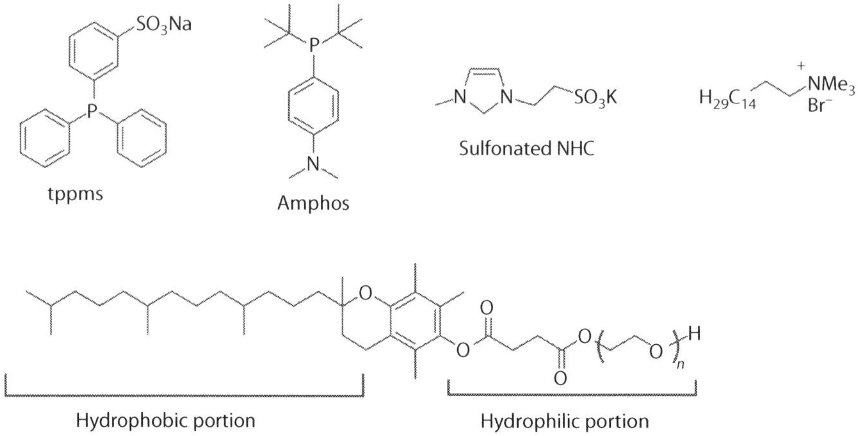

FIGURE 2.4 Some water-soluble ligands for cross-coupling reactions.

tailor-made polyethylene glycol surfactants have proven effective, providing high yields that are competitive on the research laboratory scale, with identical reactions run in traditional organic solvents (Figure 2.4, bottom).[62,63] In several examples, cross-coupling reactions performed in water/surfactant mixtures do not require any special handling (air sensitive, degassed solvents, etc.) and the product isolation is not any more labor-intensive or energy-intensive than the traditional methods of filtration or distillation.

Regardless of the solvent chosen, another hurdle in the overall process is the separation of the catalyst from the reaction mixture or desired product (or both). This is usually achieved by using heterogeneous catalysts; however, in many cases, there is a loss of selectivity due to the lack of a defined single active catalytic site. One way of overcoming this obstacle is to tether a single site catalyst to a solid substrate to form a heterogeneous catalyst (Figure 2.5). This methodology takes advantage of

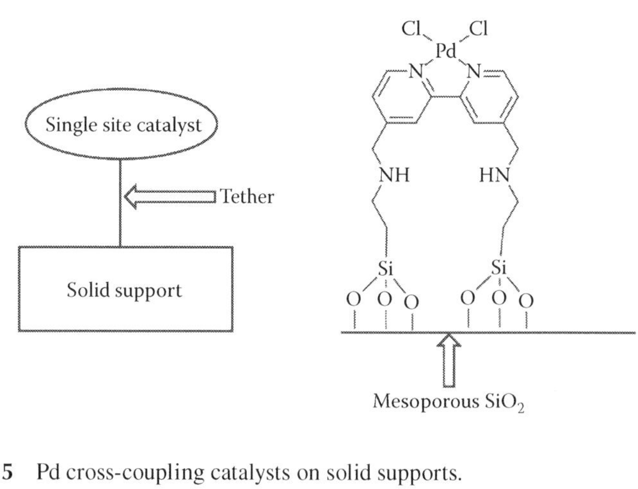

FIGURE 2.5 Pd cross-coupling catalysts on solid supports.

the ease of separation/isolation of heterogeneous catalyst materials after a reaction while maintaining the selectivity of a single-site transition metal catalyst. Several examples of solid supports have been reported in the literature and have been shown to be highly active and selective. One such example uses mesoporous silica nanoparticles (MSNs) (a specially synthesized nanoparticle) as a support for a Pd catalyst used in Heck couplings (Figure 2.5).[64] This catalyst actually performed better than the homogeneous Pd system under identical reaction conditions. It has the additional benefit of easy isolation by filtration and could be recycled four times without loss of activity. Other examples of MSN supported Pd complexes have been used with similar results for the other cross-coupling reactions.[60,65] Another very interesting area of this type is that of magnetically recyclable nanocatalyts.[66–68] Here, the catalyst is tethered to a support that is magnetic, usually one that is or contains Fe_2O_3. In this case, the catalyst material can simply be recovered from the reaction medium using a magnet or magnetic field. Although the use of nanocatalysts is both exciting and promising, the toxicity of nanoparticles in general has only recently been investigated and should be better understood before the wide-scale application and adoption of these catalysts.[69]

Finally, alternate metals such as copper and iron have been investigated as catalysts in place of Pd. The advantages of using these metals include the fact that they are earth-abundant elements, less toxic, and possibly cheaper. Fe should be preferred over Cu because of lower toxicity and relatively higher abundance. However, when compared with research using Pd, alternative metals have not been investigated in detail and industrial acceptance will require much more in-depth research. Furthermore, the use of these metals requires higher catalyst loading (5–10 mol%), and thus far examples have reduced recyclability raising the questions of possible increased overall waste and costs.[70–75]

As mentioned earlier, cross-coupling reactions are frequently used in industry. Currently, the only green aspect of these reactions is their use of catalysts. The developments mentioned previously including the use of ligand-free palladium catalysts, alternative solvents, microwave heating, etc., have not been demonstrated on an industrial scale but have thus far been largely restricted to academic studies. Part of the reason for this is that academic research seems to be fragmented, with some teams looking at alternative solvents whereas others focus on greener catalysts, for instance. It is the opinion of the authors that interdisciplinary academic research is required to develop a holistically green cross-coupling reaction keeping in mind all the stages and components of the reaction. A life cycle assessment of cross-coupling reactions, including all stages of the process such as the disposal and use of the ultimate products is needed. Such research will allow for the benefits in cost and sustainability to be evaluated readily and allow for quicker adoption by industry.

One of the common themes of the C–C bond formation mechanism mentioned previously is that they all require activated substrates and produce salt and other waste streams. To circumvent these and other problems and increase atom economy, it is necessary to imagine a completely new mechanism for the formation of carbon–carbon bonds and develop catalysts that could perform the same bond-forming reactions through a different pathway.

2.1.2 Olefin Hydroarylation and Hydroalkylation

The global production of ethylbenzene was approximately 20 million metric tons in 2002, most of which was converted into styrene and eventually polymerized.[76] Approximately 24% of the ethyl benzene produced industrially is synthesized through Friedel–Crafts (FC) alkylation using Lewis acid catalysts (e.g., $AlCl_3$ and zeolites) coupled with strong mineral acids at high temperature and pressure.[77] This process has several drawbacks including the inability to recycle the catalyst, high energy input, lack of regioselectivity, and a propensity to produce high yields of polyalkylated arenes that must undergo further transformation to produce the desired alkyl arene (Scheme 2.7). If propylene is used instead of ethylene, the FC mechanism almost exclusively produces branched products.

One new approach to the formation of C–C bonds is that of olefin hydroarylation or hydroalkylation. This mechanism has been heavily studied in recent years and operates at essentially 100% atom economy.[78] It does not require the use of any activated substrates and instead uses unsubstituted arenes and olefins as the reagents for coupling. The overall reaction involves a transition metal that is able to selectively activate/break arene C–H bonds (Scheme 2.8). This is immediately followed by a binding and

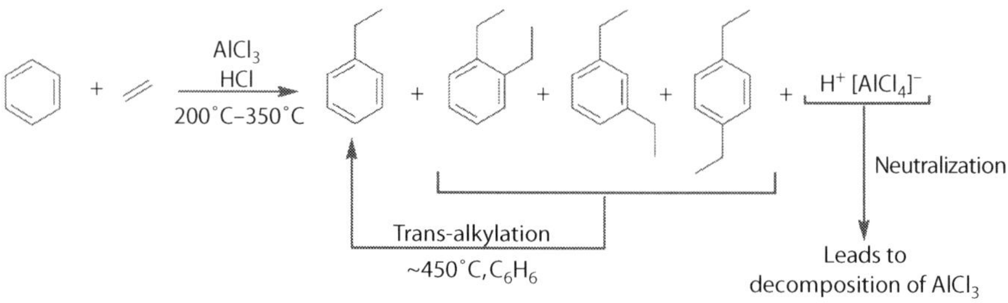

SCHEME 2.7 The production of ethylbenzene using Friedel–Crafts alkylation.

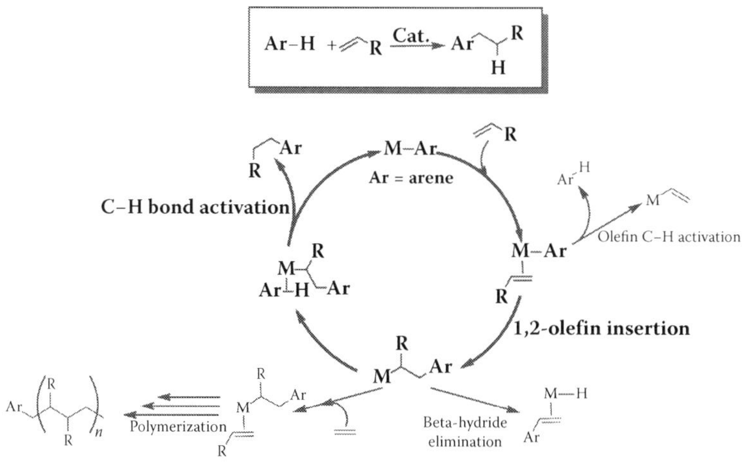

SCHEME 2.8 General olefin hydroarylation mechanism.

insertion of an olefin into the new metal-C_{arene} bond. Subsequent C–H bond activation liberates the product and returns the catalyst to its initial state to continue the cycle.

From Scheme 2.8, it can be seen that there are several potential side reactions (Olefin C–H activation, β-hydride elimination and polymerization labeled with italics) that are not part of the catalytic cycle for the production of alkylarenes. These are included to demonstrate the difficulty in finding an appropriate metal–ligand complex that is capable of selectively performing the necessary steps to transform benzene to ethylbenzene (in the simplest) case.

Current transition metal catalysts that are able to transform benzene and olefins into alkyl arenes are based on Ru, Ir, Pt, and some Rh complexes. These systems have been studied in great detail and perform well under mild conditions (<200°C, <200 psi ethylene). In addition to a 100% atom economy, the catalysis is performed neatly, meaning that no solvents are used. Benzene serves as both the substrate and the solvent in these cases. Additionally, this process is selective for monoalkylated arenes, such as ethyl benzene.

This methodology is not without its limitations. The production of nonbranched alkyl arenes (for propylene and higher olefins) is still a challenge that has yet to be met with high selectivity. Additionally, it still relies on the rare expensive metals PGMs (Ru, Pt, Ir, and Rh). There are concerns with the availability, environmental problems of mining, and overall toxicity of these metals.[79] The frontier of this area, as in that of cross-coupling, will be the development of selective catalysts based on earth-abundant metals (Fe, Mn, etc.).[80–82]

2.1.2.1 C–O Bond Formation

In contrast to the carbon–heteroatom bonds (C–O, C–N, etc.) formed through cross-coupling used to manufacture fine chemicals, C–O bonds in many commodity chemicals are synthesized through a different mechanism. Two classic reactions for the production of commodity chemicals typically come to mind with the mention of C–O bond formation: the Wacker (pronounced "Vacker") and the Monsanto acetic acid processes. These methodologies laid the groundwork for the use of single active site transition metal catalysis becoming viable for use in many industrial processes.

The Wacker process, which is used to produce approximately 6 million tons of acetaldehyde (CH_3CHO) per year, uses a combination of palladium(II) and copper(II) to oxidize ethylene to acetaldehyde in highly atom economical processes.[83] The reaction of ethylene with aqueous solution of $PdCl_2$ to stoichiometrically synthesize aldehydes had been known since the 1890s (Figure 2.6, top). The stoichiometric

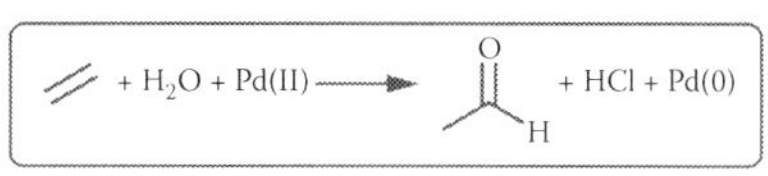

$$C_2H_2 + 0.5\,O_2 \longrightarrow CH_3C(O)H \qquad (3)$$

$$Pd(0) + 2\,ClCl_2 \longrightarrow PdCl_2 + 2\,CuCl$$
$$2\,CuCl + 2\,HCl + 0.5\,O_2 \longrightarrow 2\,CuCl_2 + H_2O$$
$$C_2H_2 + 0.5\,O_2 \longrightarrow CH_3C(O)H$$

FIGURE 2.6 General reaction for the hydroformylation of olefins.

consumption of expensive Pd rendered this reaction industrially unfeasible until the discovery, by Wacker Inc., that $CuCl_2$ could be used in conjunction with O_2 to reoxidize the Pd^0 to Pd^{II} to allow for reentry into a catalytic cycle. It is important to note that O_2 is not the O atom source in the final product. The carbonyl oxygen actually comes from water, which also serves as one of the solvents for the reaction. This changed the reaction equation above to provide an atom economical and commercially viable method to produce a commodity chemical even though it uses, in part, expensive Pd as the main active metal. The utilization of copper(II) salts and O_2 as an oxidant (as well as other systems that work similarly) has found use in a wide variety of applications, especially in academic laboratories developing new methods for all types of transformations (Figure 2.6, bottom). For example, the copper(I/II)/O_2 oxidation cycle has been successfully applied to the oxidation of methane to methanol using platinum catalysts.[84,85] The introduction of the copper(II/I) cycle running concomitantly with the hydroformylation reaction yields an overall reaction where O_2 is used indirectly as the O source in the product. The conditions for the industrial-scale Wacker process is relatively mild at 50°C to 130°C and 3 to 10 atm of ethylene. One major disadvantage of this process is the corrosive nature of the reaction due to the aqueous acidic conditions, which require reactors to be built from expensive materials, most commonly titanium alloys.[86] Additionally, all Wacker-type oxidations suffer from high concentrations of Cl^- that lead to the formation of undesired chlorinated by-products. Current work toward solving these problems (i.e., acid-free and $CuCl_2$-free solutions) focuses on the synthesis of ligated, water-soluble Pd compounds and acetamide solvents in biphasic reactions. These include changing the reaction conditions from the organic/aqueous biphasic system to one that uses ionic liquid solvents or ionic liquid/super critical CO_2 mixed solvent systems.[87]

The second of the C–O bond formation reactions, the Monsanto acetic acid process (more generally known as a carbonylation reaction), is another fundamental reaction used industrially to form C–O bonds. More than 8 million tons of acetic acid (CH_3COOH) is produced annually using a rhodium(I) catalyst to combine carbon monoxide with methanol with more than 99% selectivity.[88] This methodology, developed by Monsanto in the late 1970s, represents one of the few industrial processes that uses homogeneous catalysts. The overall reaction involves the insertion of CO into the C–O bond of methanol (CH_3OH) to form acetic acid (Figure 2.7).

Although this process proceeds at high selectivity for acetic acid formation and at relatively mild conditions (150°C–200°C and 30–60 atm CO), there are several limitations to the Rh catalyzed process: lower than 90% CO use, catalyst deactivation, and costly product isolation. One of the other requirements is the need for approximately 8% water in the reactor to ensure that the addition of CH_3I to the Rh(I) center is the rate-determining step of the cycle (Scheme 2.9). The additional water can undergo an

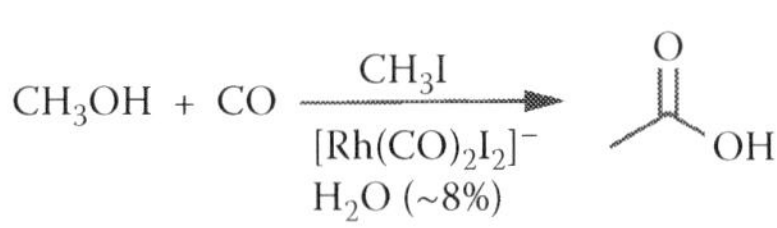

FIGURE 2.7 Carbonylation of methanol to produce acetic acid.

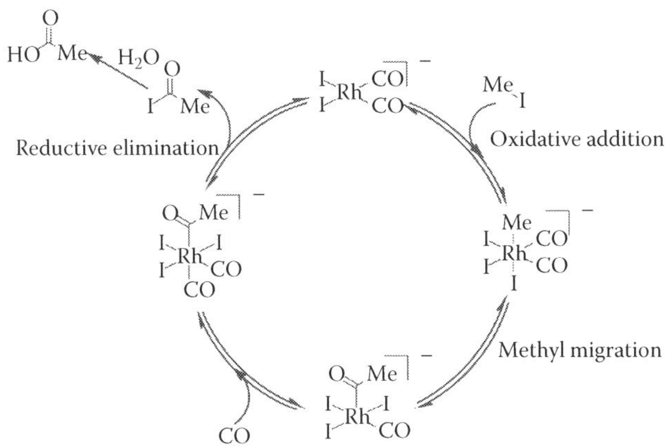

SCHEME 2.9 Mechanism for rhodium-catalyzed carbonylation.

Rh(I) catalyzed reaction with CO to form CO_2 and H_2 (called the water–gas shift reaction), which must be purged from the system and leads to a lower efficiency for the conversion of CO to acetic acid. The high cost of isolation of the acetic acid product is due not only to the need to remove the required excess water from the reaction medium but also propanoic acid, which is an additional side product of the carbonylation process.

A major improvement to this system, known as CATIVA™ and developed by BP, uses an iridium catalyst in conjunction with ruthenium, resulting in a rate increase of about 150 times that of the Rh(I) system.[89] This catalyst system is more soluble, improves reaction rates, and uses lower percentages of water leading to fewer by-products, greater use of CO (>90%), and lower purification costs.[1] Additionally, the iridium catalyst system is significantly more stable than the rhodium system, leading to longer catalyst lifetimes and decreased overall costs.[86] The improvement that this system has demonstrated has been attributed somewhat to a change in the rate-limiting step of the reaction from oxidative addition of CH_3I with Rh to the insertion of CO into the Ir-CH_3 bond in the CATIVA system. Additionally, the addition of the ruthenium promoter complex results in a 150-fold increase in the rate-limiting CO insertion at the Ir center.[90]

2.1.2.2 Hydroformylation of Alkenes

An additional large-scale use of homogeneous transition metal catalysts involves the production of aldehydes through the hydroformylation of olefins, also known as the oxo-process (Figure 2.8). For example, more than 14 billion pounds of aldehydes are produced per year using this 100% atom economic process; this amounts to just under 2 pounds of aldehyde for each person on Earth per year.[15] Aldehydes are used in a variety of applications such as pharmaceuticals, polymers, materials, and many

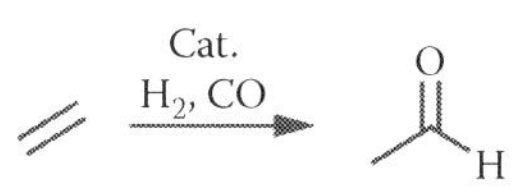

FIGURE 2.8 General reaction for the formation of aldehydes through the oxo-process.

others. For lower molecular weight olefins (most of which are gases at room temperature), a straightforward distillation of the low-boiling aldehyde product makes for an inexpensive isolation from the catalyst.[7] This, however, limits the application of this process to higher olefins that could be used in the direct production of pharmaceuticals, with the notable example of one of the steps in the synthesis of vitamin A.[91] Hydroformylation using an Rh catalyst is applied on the more than 100 g scale in the synthesis of fragrances.[92]

Whether using Rh or Co catalysts, the pathway for the hydroformylation of olefins is essentially the same. The catalytic cycle is initiated through the formation of a metal hydride (M-H) followed by binding of an η^2-olefin. Insertion into the M-H bond to yield the metal alkyl complex is followed by binding of CO and subsequent migration of the M-C_{alkyl} bond to yield the metal acyl complex. The activation of H_2 completes the cycle to facilitate the release of product and to regenerate the transition metal hydride to reenter the catalytic cycle (Scheme 2.10). The original work by Otto Roelen in the late 1930s used $Co_2(CO)_8$ as a precursor that breaks into two $Co(CO)_4H$ units upon exposure to H_2 and requires the dissociation of a CO ligand before olefin binding. The other workhorse of industrial hydroformylation usually involves a phosphine ligated Rh(I) complex of the general formula $HRh(CO)_2(PPh_3)_2$.

One of the drawbacks of this process is the potential formation of two regioproducts (Figure 2.9). The insertion of the olefin into the M-H bond can occur in two ways. One possible pathway occurs when the olefin undergoes a Markovnikov-type addition and the hydride ends up on the less substituted vinylic carbon (leading to the 2° or branched product) or the olefin can insert (as shown in Scheme 2.10), where the more substituted carbon atom receives the hydride, which leads to the less substituted linear (1°) product (Scheme 2.7). The competition for 1° (linear/terminal) versus 2° (branched) product formation selectivity results not from the olefin insertion step (which is reversible), but from the irreversible migratory insertion step of

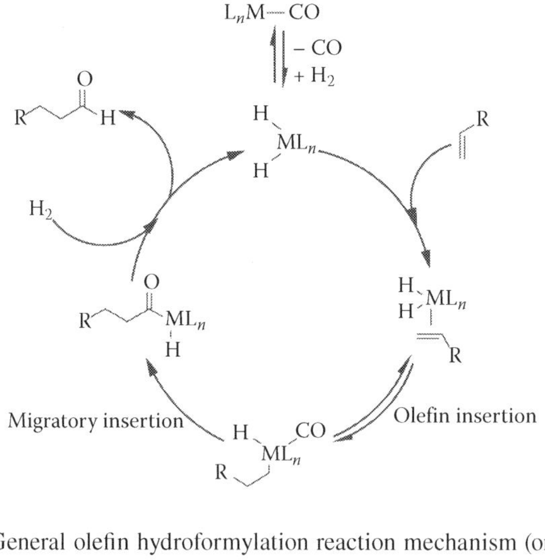

SCHEME 2.10 General olefin hydroformylation reaction mechanism (only 1° product formation shown).

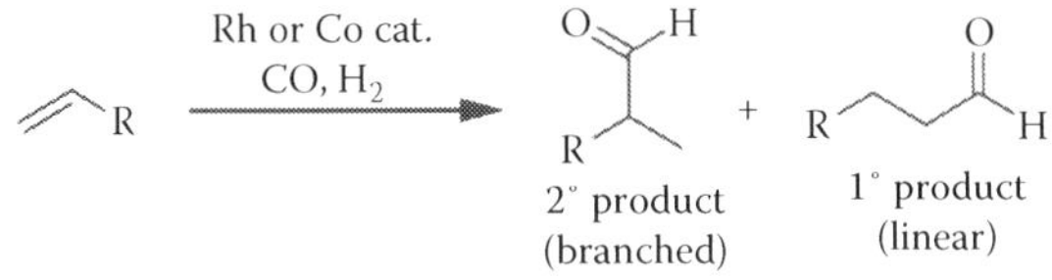

FIGURE 2.9 Production of linear and branched aldehyde products.

the catalytic cycle. For the Co catalyzed system, the addition of a bulky phosphine ligand, P(*n*-butyl)$_3$, not only increased the activity of the catalyst, but also shows a preference for the 1° aldehyde product by about 8:1 versus 4:1 in the original system. By switching to the more expensive Rh phosphine complexes, the rate was again accelerated (1 atm CO/H_2 and 25°C) and an even higher selectivity for the 1° aldehyde product. Additionally, both Co and Rh are active catalysts for alkene isomerization, which means that almost the same aldehyde product distributions can be obtained using 2-butene as the starting material versus 1-butene. This implies that 1° *n*-aldehydes can be synthesized from internal alkenes that are usually significantly cheaper than their terminal alkene counterparts. For example, 1-pentene is approximately 30% more expensive than 2-pentene per gram.*

Hydroformylation in water has been thoroughly studied and has been practiced on the industrial scale since the mid-1980s. While at Rhone–Poulenc, Kuntz began to study an Rh system that replaced the usually non–water-soluble phosphine ligands (PPh_3) with sulfonated phosphine ligands ($P(3\text{-}C_6H_4SO_3Na)_3$) known as the Ruhrchemie/Rhone–Poulenc process.[93] This biphasic reaction system is used to produce millions of tons of low molecular weight aldehydes per year. This Rh-based system has the following advantages:[87]

- Enhanced catalyst activity and selectivity for linear aldehyde production
- Efficient and easy catalyst recovery
- Higher energy efficiency (process is actually a net heat producer) and near total elimination of harmful plant emissions

The major limitation, as with other hydroformylation systems, is the incompatibility with higher molecular weight olefins (>C_5) due to their poor solubility in water and limited mass transport.

Current studies, as with many other processes, are focused on the switch from the earth rare and expensive Rh-based systems to alternative metals.[94] There are several systems based on Ru, Ir, Pd, and Pt, which are also in high demand and relatively earth rare. By far the most promising is the development of Fe-based systems that have shown limited success. Future work in the area will include a comprehensive sampling of different ligands to improve the selectivity and activity of Fe-based hydroformylation systems.

* Calculated from the price for a 25 g bottle of >98% purity from sigmaaldrich.com on December 10, 2014.

2.2 CONCLUSION

The development of new transition metal complexes is still at the forefront of studies in the inorganic and organometallic community. The future lies in not just the development of new catalysts or ligands, but in working in conjunction with the green chemistry community and engineers to develop new solvent systems or purification techniques/mechanisms. Currently, most of the chemicals we encounter in everyday life have their origin in crude oil. Crude oil is a complex mixture, which gets broken down and built back up; it seems a convoluted way of going about things. Highly efficient catalytic C–C bond building reactions will also allow the use of simple molecules like CO_2 and merely build up molecules without going through the breaking down and rebuilding phases. Additionally, the use of waste products such as CH_4 (from oil drilling) or CO_2 (from just about any industrial process or car, etc.) as feedstock will play an essential role in lowering the cost and, to a small extent, the environmental impact of the production of the essentials for everyday life.

REFERENCES

1. Anastas, P.T., and Warner, J.C. *Green Chemistry: Theory and Practice*. Oxford University Press, New York: 2000; 275–277.
2. Anastas, P.T., and Warner, J.C. *Green Chemistry: Theory and Practice*. Oxford University Press, New York: 2000; 315.
3. Steinborn, D. *Fundamentals of Organometallic Catalysis*. Wiley-VCH, Weinheim, Germany: 2012.
4. Heaton, B. *Mechanisms in Homogeneous Catalysis: A Spectroscopic Approach*. Wiley-VCH, Weinheim, Germany: 2006.
5. van Leeuwen, P.W.N.M. *Homogeneous Catalysis; Understanding the Art*. Kluwer Academic Publishers, Dordrecht, the Netherlands (now Springer): 2004.
6. Anastas, P., and Eghbali, N. *Chem. Soc. Rev.* 2010, 39, 301–312.
7. Lancaster, M., and Lancaster, M. *Green Chemistry: An Introductory Text*. Royal Society of Chemistry, New York, New York: 2010; 118–121.
8. Johansson Seechurn, C.C.C., Kitching, M.O., Colacot, T.J., and Snieckus, V. *Angew. Chem. Int. Ed.* 2012, 51, 5062–5085.
9. Han, F.-S. *Chem. Soc. Rev.* 2013, 42, 5270–5298.
10. Kambe, N., Iwasaki, T., and Terao, J. *Chem. Soc. Rev.* 2011, 40, 4937–4947.
11. Diederich, F., and Stang, P.J. *Metal-Catalyzed Cross-Coupling Reactions*. Wiley-VCH, Weinheim, Germany: 2008.
12. Corbet, J.-P., and Mignani, G. *Chem. Rev.* 2006, 106, 2651–2710.
13. Magano, J., and Dunetz, J.R. *Chem. Rev.* 2011, 111, 2177–2250.
14. Heck, R.F., Negishi, E., and Suzuki, A. *Scientific Background of the Nobel Prize in Chemistry 2010. Palladium-Catalyzed Cross Couplings in Organic Synthesis*. The Royal Swedish Academy of Sciences, Stockholm, Sweden: 2010.
15. Hartwig, J.F. *Organotransition Metal Chemistry: From Bonding to Catalysis*. University Science Books, Sausalito, CA: 2010; 751–752.
16. Hartwig, J.F. *Acc. Chem. Res.* 2008, 41, 1534–1544.
17. Martin, R., and Buchwald, S.L. *Acc. Chem. Res.* 2008, 41, 1461–1473.
18. Negishi, E., and Anastasia, L. *Chem. Rev.* 2003, 103, 1979–2018.
19. Blaser, H., Indolese, A., Naud, F., Nettekoven, U., and Schnyder, A. *Adv. Synth. Catal.* 2004, 346, 1583–1598.

20. Littke, A.F., and Fu, G.C. *Angew. Chem. Int. Ed.* 2002, 41, 4176–4211.
21. Crabtree, R.H. *The Organometallic Chemistry of the Transition Metals*, 6th ed. John Wiley & Sons, Hoboken, NJ: 2014; 249.
22. Li, Z., Jiang, Y., and Fu, Y. *Chem.—Eur. J.* 2012, 18, 4345–4357.
23. Saito, B., and Fu, G.C. *J. Am. Chem. Soc.* 2008, 130, 6694–6695.
24. Lundin, P.M., and Fu, G.C. *J. Am. Chem. Soc.* 2010, 132, 11027–11029.
25. Crudden, C.M., Glasspoole, B.W., and Lata, C.J. *Chem. Commun.* 2009, 6704–6716.
26. Rudolph, A., and Lautens, M. *Angew. Chem. Int. Ed.* 2009, 48, 2656–2670.
27. Cordier, C.J., Lundgren, R.J., and Fu, G.C. *J. Am. Chem. Soc.* 2013, 135, 10946–10949.
28. Crabtree, R.H. *The Organometallic Chemistry of the Transition Metals*, 6th ed. John Wiley & Sons, Hoboken, NJ: 2014; 248.
29. Colacot, T.J. *Platin. Met. Rev.* 2011, 55, 84–90.
30. Miyaura, N., and Suzuki, A. *Chem. Rev.* 1995, 95, 2457–2483.
31. Kündig, E.P., Jia, Y., Katayev, D., and Nakanishi, M. *Pure Appl. Chem.* 2012, 84, 1741–1748.
32. Li, H., Johansson Seechurn, C.C.C., and Colacot, T.J. *ACS Catal.* 2012, 2, 1147–1164.
33. Hoffmann, R. *Frontiers in Chemstry.* Pergamon, Oxford: 1982; 247–263.
34. de Vries, J.G. *Can. J. Chem.* 2001, 79, 1086–1092.
35. Kotha, S., Lahiri, K., and Kashinath, D. *Tetrahedron* 2002, 58, 9633–9695.
36. Nicolaou, K.C., Bulger, P.G., and Sarlah, D. *Angew. Chem. Int. Ed.* 2005, 44, 4442–4489.
37. Overman, L.E., Ricca, D.J., and Tran, V.D. *J. Am. Chem. Soc.* 1993, 115, 2042–2044.
38. Wipf, P., and Lim, S. *J. Am. Chem. Soc.* 1995, 117, 558–559.
39. Oh, C.H., and Jung, S.H. *Tetrahedron Lett.* 2000, 41, 8513–8516.
40. Shaughnessy, K., and Greener, H. Approaches to cross-coupling in new trends in cross-coupling: Theory and applications; Colacot, T. (Ed.); *RSC Catalysis Series No. 21.* The Royal Society of Chemistry, Cambridge, UK: 2014; 645–696.
41. Capello, C., Fischer, U., and Hungerbuhler, K. *Green Chem.* 2007, 9, 927–934.
42. Clark, J.H., and Tavener, S.J. *Org. Process Res. Dev.* 2007, 11, 149–155.
43. Henderson, R.K., Jimenez-Gonzalez, C., Constable, D.J.C., Alston, S.R., Inglis, G.G.A., Fisher, G., Sherwood, J., Binks, S.P., and Curzons, A.D. *Green Chem.* 2011, 13, 854–862.
44. Vallin, K.S.A., Emilsson, P., Larhed, M., and Hallberg, A. *J. Org. Chem.* 2002, 67, 6243–6246.
45. Dilip, M. *Nanomater. Energy* 2012, 1, 193–206.
46. Plechkova, N.V., and Seddon, K.R. *Chem. Soc. Rev.* 2008, 37, 123–150.
47. Polshettiwar, V., Decottignies, A., Len, C., and Fihri, A. *ChemSusChem* 2010, 3, 502–522.
48. Cave, G.W.V., Raston, C.L., and Scott, J.L. *Chem. Commun.* 2001, 2159–2169.
49. Liang, Y., Xie, Y.-X., and Li, J.-H. *J. Org. Chem.* 2006, 71, 379–381.
50. Botella, L., and Najera, C. *Angew. Chem. Int. Ed.* 2002, 41, 179–181.
51. Wan, J.-P., Wang, C., Zhou, R., and Liu, Y. *RSC Adv.* 2012, 2, 8789–8792.
52. Wolfson, A., and Dlugy, C. *Chem. Pap.* 2007, 61, 228–232.
53. Čapek, P., Pohl, R., and Hocek, M. *Org. Biomol. Chem.* 2006, 4, 2278–2284.
54. Kostas, I.D., Heropoulos, G.A., Kovala-Demertzi, D., Yadav, P.N., Jasinski, J.P., Demertzis, M.A., Andreadaki, F.J., Vo-Thanh, G., Petit, A., and Loupy, A. *Tetrahedron Lett.* 2006, 47, 4403–4407.
55. Lépine, R., and Zhu, J. *Org. Lett.* 2005, 7, 2981–2984.
56. Dallinger, D., and Kappe, C.O. *Chem. Rev.* 2007, 107, 2563–2591.
57. Diéguez, M., Pàmies, O., Ruiz, A., Diaz, Y., Castillón, S., and Claver, C. *Coord. Chem. Rev.* 2004, 248, 2165–2192.
58. Godoy, F., Segarra, C., Poyatos, M., and Peris, E. *Organometallics* 2011, 30, 684–688.

59. Lipshutz, B., Abela, A., Boskovic, Z.V., Nishikata, T., Duplais, C., and Krasovskiy, A. *Top Catal.* 2010, 53, 985–990.
60. Lipshutz, B.H., Taft, B.R., Abela, A.R., Ghorai, S., Krasovskiy, A., and Duplais, C. *Platin. Met. Rev.* 2012, 56, 62–74.
61. Lipshutz, B.H., Ghorai, S., Abela, A.R., Moser, R., Nishikata, T., Duplais, C., Krasovskiy, A., Gaston, R.D., and Gadwood, R.C. *J. Org. Chem.* 2011, 76, 4379–4391.
62. Lipshutz, B.H., and Ghorai, S. *Aldrichim. Acta* 2012, 45, 3–16.
63. La Sorella, G., Strukul, G., and Scarso, A. *Green Chem.* 2015, 17, 644–683.
64. Tsai, F.-Y., Wu, C.-L., Mou, C.-Y., Chao, M.-C., Lin, H.-P., and Liu, S.-T. *Tetrahedron Lett.* 2004, 45, 7503–7506.
65. Sharma, K.K., Biradar, A.V., Das, S., and Asefa, T. *Eur. J. Inorg. Chem.* 2011, 2011, 3174–3182.
66. Dwars, T., Paetzold, E., and Oehme, G. *Angew. Chem. Int. Ed.* 2005, 44, 7174–7199.
67. Gawande, M.B., Luque, R., and Zboril, R. *ChemCatChem* 2014, 6, 3312–3313.
68. Wang, D., Deraedt, C., Salmon, L., Labrugre, C., Etienne, L., Ruiz, J., and Astruc, D. *Chem. Eur. J.* 2015, 21, 1508–1519.
69. Sharifi, S., Behzadi, S., Laurent, S., Forrest, M.L., Stroeve, P., and Mahmoudi, M. *Chem. Soc. Rev.* 2012, 41, 2323–2343.
70. Monnier, F., and Taillefer, M. *Angew. Chem. Int. Ed.* 2009, 48, 6954–6971.
71. Surry, D.S., and Buchwald, S.L. *Chem. Sci.* 2010, 1, 13–31.
72. Beletskaya, I.P., and Cheprakov, A.V. *Organometallics* 2012, 31, 7753–7808.
73. Beletskaya, I.P., and Cheprakov, A.V. *Coord. Chem. Rev.* 2004, 248, 2337–2364.
74. Buchwald, S.L., and Bolm, C. *Angew. Chem. Int. Ed.* 2009, 48, 5586–5587.
75. Correa, A., Mancheño, O.G., and Bolm, C. *Chem. Soc. Rev.* 2008, 37, 1108–1117.
76. Perego, C., and Ingallina, P. *Catal. Today* 2002, 73, 3–22.
77. Olah, G.A., and Molnar, A. *Hydrocarbon Chemistry.* John Wiley & Sons, New York: 2003.
78. Andreatta, J.R., McKeown, B.A., and Gunnoe, T.B. *J. Organomet. Chem.* 2011, 696, 305–315.
79. Jha, M.K., Lee, J., Kim, M., Jeong, J., Kim, B.-S., and Kumar, V. *Hydrometallurgy* 2013, 133, 23–32.
80. Kischel, J., Jovel, I., Mertins, K., Zapf, A., and Beller, M. *Org. Lett.* 2006, 8, 19–22.
81. Kalman, S.E., Petit, A., Gunnoe, T.B., Ess, D.H., Cundari, T.R., and Sabat, M. *Organometallics* 2013, 32, 1797–1806.
82. Ilies, L., and Nakamura, E. *Iron-Catalyzed Cross-Coupling Reactions.* John Wiley & Sons, Ltd., New York: 2009.
83. Imandi, V., Kunnikuruvan, S., and Nair, N.N. *Chem. Eur. J.* 2013, 19, 4724–4731.
84. Muehlhofer, M., Strassner, T., and Herrmann, W.A. *Angew. Chem. Int. Ed.* 2002, 41, 1745–1747.
85. Periana, R.A., Taube, D.J., Gamble, S., Taube, H., Satoh, T., and Fujii, H. *Science* 1998, 280, 560–564.
86. Sheldon, R.A., Arends, I., and Hanefeld, U. *Green Chemistry and Catalysis.* John Wiley & Sons, New York: 2007; 247.
87. Sheldon, R.A., Arends, I., and Hanefeld, U. *Green Chemistry and Catalysis.* John Wiley & Sons, New York: 2007; 302–304.
88. Omae, I. *Appl. Organomet. Chem.* 2009, 23, 91–107.
89. Winterton, N. *Chemistry for Sustainable Technologies: A Foundation.* Royal Society of Chemistry, London: 2011; 339–343.
90. Haynes, A., Maitlis, P.M., Morris, G.E., Sunley, G.J., Adams, H., Badger, P.W., Bowers, C.M., Cook, D.B., Elliott, P.I.P., and Ghaffar, T. *J. Am. Chem. Soc.* 2004, 126, 2847–2861.

91. Averill, B.A., Moulijn, J.A., van Senten, R.A., and van Leeuwen, P.W.N.M. *Catalysis: An Integrated Approach*. Elsevier, Armsterdam, the Netherlands: 1999.
92. Whiteker, G.T., and Cobley, C.J. *Applications of Rhodium-Catalyzed Hydroformylation in the Pharmaceutical, Agrochemical, and Fragrance Industries*. Springer, Berlin, Heidelberg: 2012; 35–46.
93. Cornils, B., and Herrmann, W.A. *Aqueous-Phase Organometallic Catalysis: Concepts and Applications*. John Wiley & Sons, New York: 2006; 351–375.
94. Pospech, J., Fleischer, I., Franke, R., Buchholz, S., and Beller, M. *Angew. Chem. Int. Ed.* 2013, 52, 2852–2872.

3 Application of Organocatalysis in Sustainable Synthesis

Craig Jamieson and Allan J. B. Watson

CONTENTS

3.1 ORGANOCATALYSIS AND SUSTAINABLE SYNTHESIS

Organocatalysis is the third principal category of catalysis following metal-based catalysis and biocatalysis. Coined by MacMillan in 2000,[1] the term *organocatalysis* brought a collective identity to an emerging field of catalysis in which bond formations were accomplished using catalytic quantities of small organic molecules. Prior to 2000, there were sporadic reports of organocatalytic transformations (which, of course, could only be classified as being organocatalytic retrospectively) (for examples, see Refs. 2–18); however, following the independent and simultaneous seminal discoveries of iminium catalysis by MacMillan[1] and enamine catalysis by Barbas, Lerner, and List[19] in 2000, reports of new organocatalytic bond formations increased exponentially. Further demonstrations of the utility of the newly discovered iminium and enamine catalysis manifolds, as well as the discovery of new organocatalytic methods of molecule activation, fuelled a singularly rapid expansion of the field (for selected books and reviews, see Refs. 20–27).

At approximately the same time that organocatalysis was emerging as a field, the principles of green chemistry were gaining traction. Although process chemists had been advocating resource and waste management for decades, the need for the entire chemical community to begin considering the sustainability of all operations, and the requirement for authoritative guidance to facilitate the transition to more favorable processes, was only just becoming apparent. Recognition of the importance of sustainability in the wider chemical community (including industry and academia) reached a tipping point at around the turn of the century with the establishment of the US Presidential Green Chemistry Challenge in 1995, the Working Group on Green Chemistry in 1996, and the Green Chemistry Institute in 1997. In addition, to formalize the philosophy and to guide research efforts in this area, a series of texts and journals began to emerge: Anastas and Warner published their seminal text "*Green Chemistry: Theory and Practice*" in 1998[28] (for selected additional texts, see Refs. 29–31) with the first dedicated journals *Clean Technologies and Environmental Policy* (Springer-Verlag) and *Green Chemistry* (Royal Society of Chemistry) published in 1998 and 1999, respectively. The issue of sustainability has since been of continually increasing importance with a series of new journals and texts on the subject emerging regularly.

Despite both fields burgeoning in the same time frame, being such a new field of research for chemical synthesis and considering the rate of its development and expansion, sustainability was not necessarily the foremost concern for organocatalysis. Consequently, as with many contemporary methodologies, improvements are required and, without further refinement (as discussed later), certain organocatalytic transformations are perhaps less likely to ever be used on any significant scale due to sustainability considerations or other practical issues. However, it should be stressed that the field of organocatalysis is not intrinsically misaligned with green chemistry principles and there are aspects that are very coherent: a general advantage is the reduced reliance on potentially problematic (e.g., toxicity and scarcity) metallic additives or catalysts. In this latter context, there are features of much more mature fields, such as metal-mediated cross-coupling, that also require improvements in relation to sustainability and, as a consequence, many of these reactions are unfavorable or are

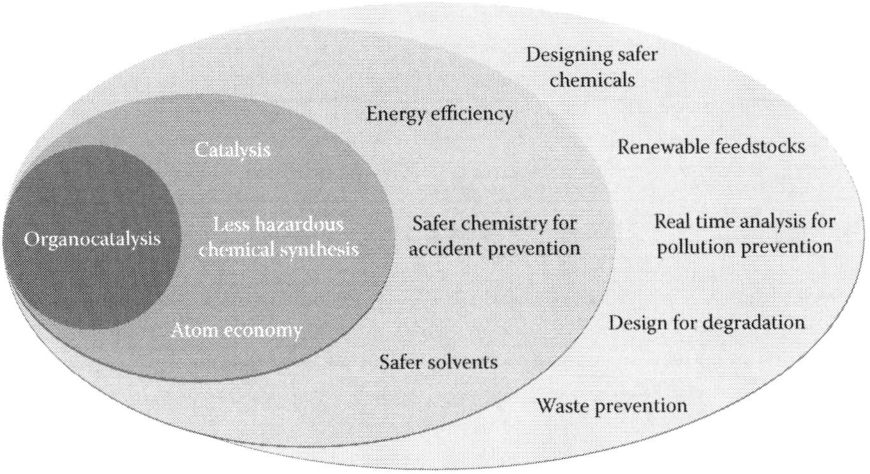

FIGURE 3.1 Coherence of organocatalysis with the Twelve Principles of Green Chemistry.

no longer used. Indeed, as the field of organocatalysis has matured simultaneously with green chemistry over the past 15 years, sustainable principles have become embedded, with some significant successes already noted. In terms of alignment with the Twelve Principles of Green Chemistry,[32] organocatalysis certainly addresses several of these; in particular, it is catalytic (principle no. 9), the majority of organocatalytic processes offer a less hazardous approach to chemical synthesis (principle no. 3), and atom economies are typically high (principle no. 2). It should be noted that certain organocatalytic processes are better aligned with green chemistry principles than others. An illustration of the coherence of the general field of organocatalysis with the Twelve Principles of Green Chemistry is shown in Figure 3.1.

Although organocatalysis has become synonymous with asymmetric catalysis, it should be noted that a wide range of nonasymmetric transformations have been extensively developed including Baylis–Hillman reactions (e.g., with DABCO), esterifications (e.g., with DMAP), Knoevenagel condensations (e.g., with piperidine), and Stetter reactions (e.g., with thiazolium salts) among many more.[20–27]

This chapter describes the application of organocatalysis to achieve enantioselective bond formations in the context of sustainable synthesis. The field of organocatalysis is vast and a full discussion of all the activation modes and developments within these is beyond the scope of this overview.[20–27] However, included in this chapter is a brief outline of the principal modes of organocatalytic activation with a concise summary of the typical reaction classes associated with each of these. General sustainability aspects of current organocatalytic technologies are then discussed.

3.2 GENERAL BRANCHES OF ORGANOCATALYSIS

Organocatalysis has seen one of the most concerted expansions of research and development in the history of chemical research. This has led to an enormous resource for practitioners of organic chemistry, and has provided solutions to some of the most challenging problems of target synthesis. Many organocatalytic reactions do not have a metal-catalyzed counterpart and have therefore facilitated access to regions of chemical space that were previously inaccessible. The field of organocatalysis can be divided into two main branches: Lewis acids/bases and Brønsted acids/bases. The branches include further series of distinct activation modes based on the general process achieved with a particular catalyst type.[20–27] A further array of possible transformations and applications exists within each of these activation modes. A hierarchical classification of the main organocatalytic processes is shown in Figure 3.2 with several examples of typical catalyst structures given in Figure 3.3 (a more detailed list of catalyst structures is given in Section 3.3.3.2).

A brief survey of each of these main branches and subbranches is given in the following sections. It should be noted that only the primary activation modes illustrated above have been discussed: there are a series of further, secondary reaction modes, which are not recorded or discussed in this overview but nevertheless add additional color to the overarching field.[20–27] Moreover, more recent developments in the field where organocatalytic activation has been merged with metal-mediated processes, such as photoredox catalysis, have not been included (for selected examples of recent developments in organocatalysis, see Refs. 33–59).

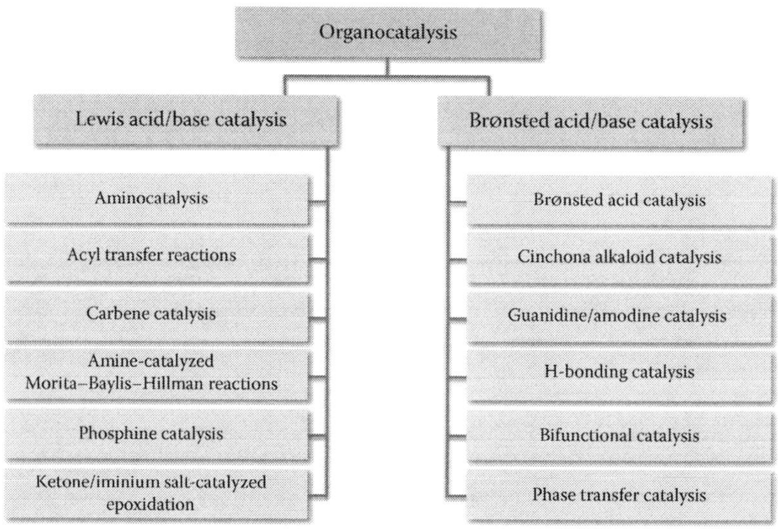

FIGURE 3.2 The main branches and subbranches of organocatalysis.

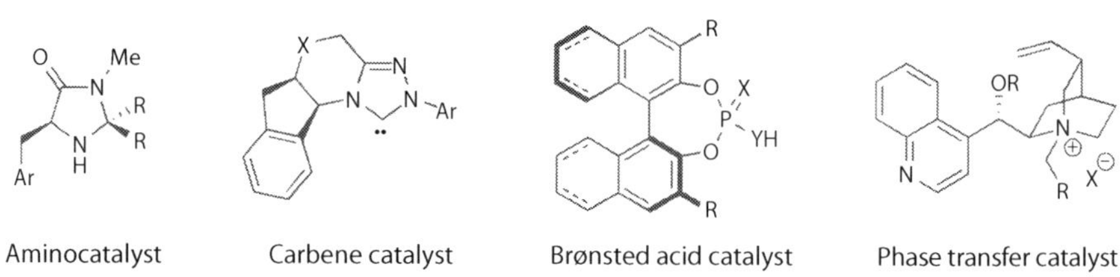

FIGURE 3.3 Examples of organocatalyst structural types.

3.2.1 Lewis Acid and Lewis Base Catalysts

3.2.1.1 Aminocatalysis: Iminium and Enamine Activation

Aminocatalysis comprises two distinct, complementary reactivity platforms that serve to functionalize carbonyl compounds: enamine and iminium catalysis[20–27] (for selected reviews, see Refs. 60–71). Enamine catalysis enables the stereoselective α-functionalization of carbonyl compounds (aldehydes and ketones). Iminium catalysis provides a method for the stereoselective β-functionalization of α,β-unsaturated carbonyl compounds (enals and enones) giving the corresponding functionalized reduced compound. Together, these orthogonal activation modes provide a raft of broadly useful and highly selective protocols for the functionalization of carbonyl compounds to deliver a wide range of synthetically valuable architectures, often in enantio-enriched form.

Activation of a suitable carbonyl compound through enamine catalysis provides a catalytic nucleophilic species capable of engaging a wide variety of electrophilic reagents, subsequently leading to an array of α-functionalized carbonyl compounds.[20–27,60–67] Enamine catalysis provides a direct, catalytic, and stereoselective alternative to established methodologies that involve preactivated carbonyl compounds, such as enolates or enol silanes. In addition, the enamine catalysis platform has seen considerable expansion beyond the use of neutral electrophilic species through the development of protocols in which electrophilic radicals are generated either through stoichiometric (singly occupied molecular orbital [SOMO] catalysis) or catalytic (photoredox catalysis) oxidants. This has substantially increased the scope of electrophiles to generate further sets of valuable transformations that were previously inaccessible through closed shell pathways. A summary of typical enamine reaction classes is provided in Figure 3.4.

Activation of α,β-unsaturated carbonyl compounds through iminium catalysis provides an alternative and complementary method to conventional metal-based Lewis acid catalysis and, indeed, provides a metal-free alternative to many of the transformations that had previously been conducted under this latter classic catalysis mode.[68–71] Moreover, iminium catalysis can provide solutions to issues previously encountered by conventional Lewis acid methodology, such as selectivity in reactions involving enones, in which lone pair discrimination has historically been challenging.[72] An illustrative summary of typical iminium-based reactions is provided in Figure 3.5.

In addition to the exemplar reaction classes described above for the individual enamine and iminium activation modes, further advances continue to be made: organocascade catalysis, achieved through the merger of iminium and enamine reactions, has produced an elegant method for the production of highly functionalized molecules, following a series of individual catalytic events, in a one-pot operation.[20–27,33–59] Initial examples focused on linking one iminium-based event with one enamine-based event but this has subsequently been extended to encompass multiple stereo-, catalyst-, and substrate-controlled events for the rapid synthesis of molecules with a rich topography. Vinylogous variants of existing enamine and iminium reactions enable stereoselective functionalization on locales that are increasingly further removed from the activating carbonyl motif[20–27] (for reviews, see Refs. 73–79). The

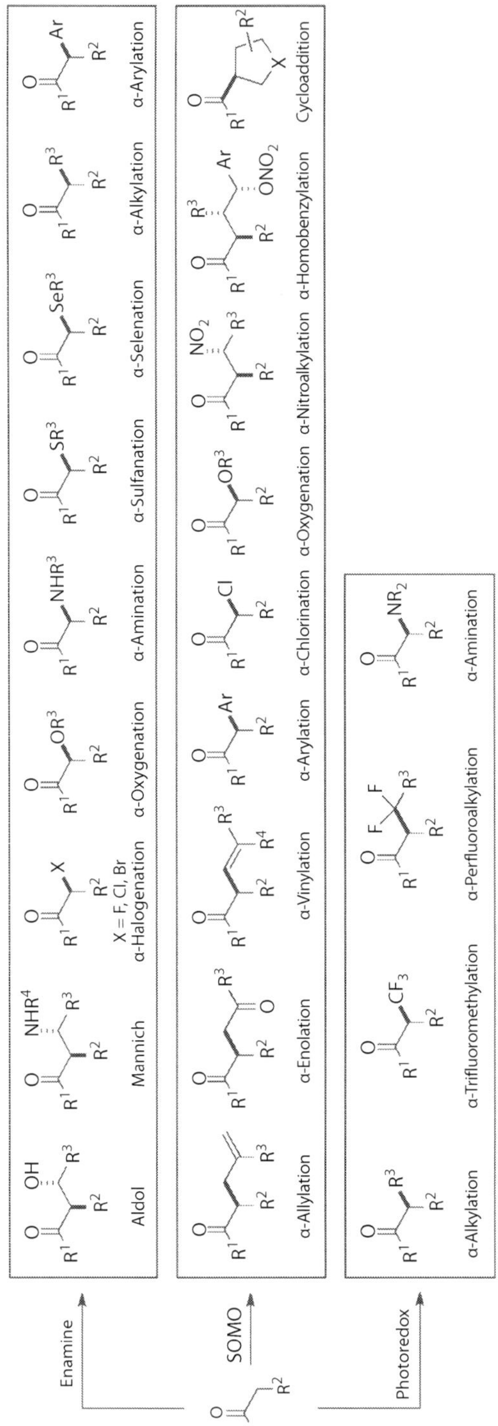

FIGURE 3.4 Illustrative summary of enamine catalysis reaction classes.

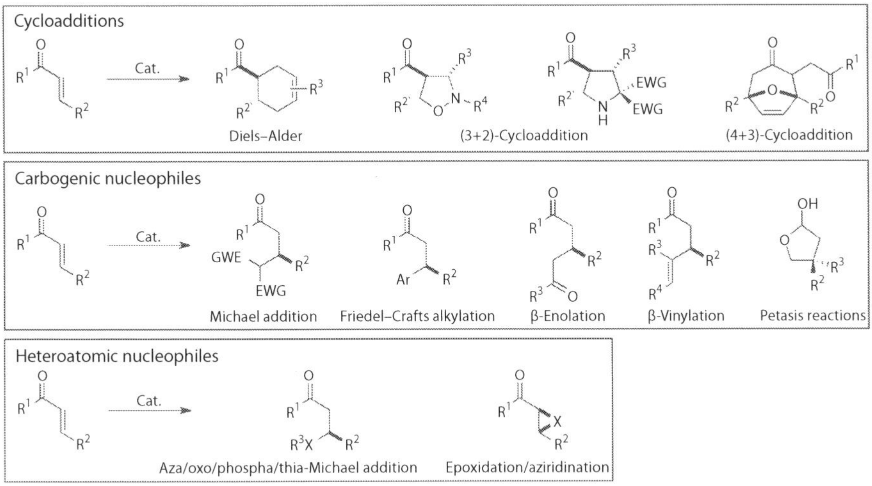

FIGURE 3.5 Illustrative summary of iminium catalysis reaction classes.

enamine-based photoredox catalysis manifold outlined in Figure 3.4 represents one example of a dual catalytic reaction platform, an area that has proven to be a fertile research topic with an array of possible synergistic catalyst combinations. Additional examples that have emerged in this area include the merger of enamine catalysis with copper catalysis and palladium catalysis to provide hitherto unprecedented reaction classes.[33–59]

3.2.1.2 Acyl Transfer Reactions

Chiral alcohols and amines are broadly employed within organic synthesis both as reagents and as chiral building blocks for pharmaceutical agents. Gaining stereoselective access to these motifs has inspired a variety of, primarily, asymmetric reduction protocols. Acyl transfer is a useful method for the preparation of enantio-enriched acylated products from racemic, *meso*, or achiral precursors (for selected reviews, see Refs. 80–84). This reaction manifold is heavily employed for the kinetic resolution of racemic alcohols and amines or in the desymmetrization of *meso* diols and diamines through selective heteroatom acylation. In addition, acyl transfer enables stereoselective *C*-acylation of achiral enol silanes or enol esters generating a variety of enantio-enriched heterocyclic scaffolds. A summary of typical acyl transfer processes is provided in Figure 3.6.

3.2.1.3 *N*-Heterocyclic Carbene Catalysis

The reaction of an *N*-heterocyclic carbene (NHC) with a suitable carbonyl compound can give rise to several reactive intermediates, which are capable of participating in a range of reactions with a variety of coupling partners and, in many cases, affords stereochemically enriched products (for selected reviews, see Refs. 85–89). Depending on the catalyst and carbonyl compound selected, NHC catalysis enables a highly versatile range of reaction types, with corresponding diversity in product

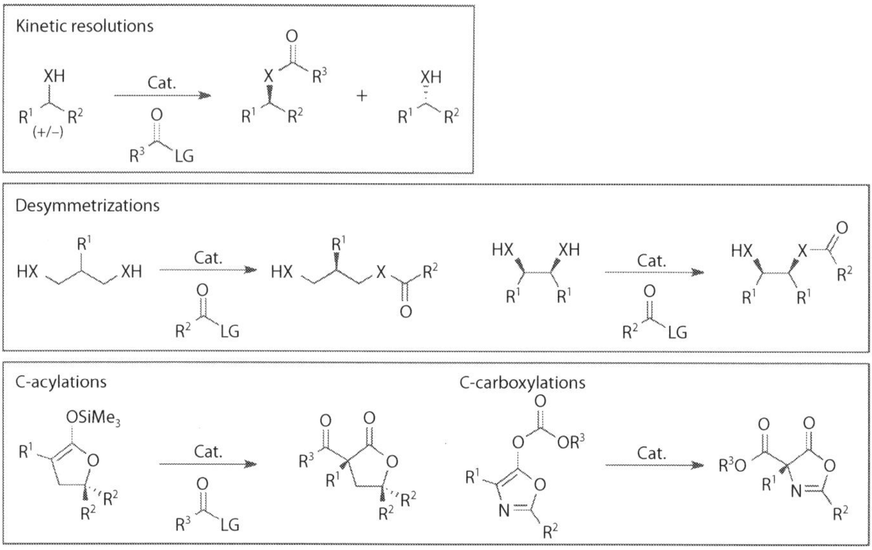

FIGURE 3.6 Illustrative summary of acyl transfer reactions.

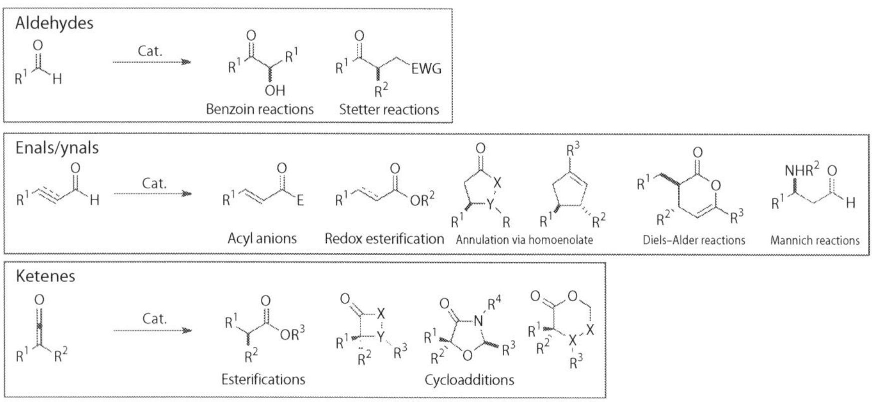

FIGURE 3.7 Illustrative summary of carbene catalysis reaction classes.

architecture. NHCs are also amenable to cascade-type processes, enabling efficient generation of highly functionalized templates for a raft of downstream applications. A summary of typical reaction classes for NHC catalysis is given in Figure 3.7.

3.2.1.4 Amine-Catalyzed Morita–Baylis–Hillman Reactions

Retrospectively classed as an organocatalytic process, the Morita–Baylis–Hillman reaction provides access to products that are similar to those obtainable through the aldol or Mannich reactions with the key difference that the nucleophile is derived from an α,β-unsaturated carbonyl compound (for selected reviews, see Refs. 90 and

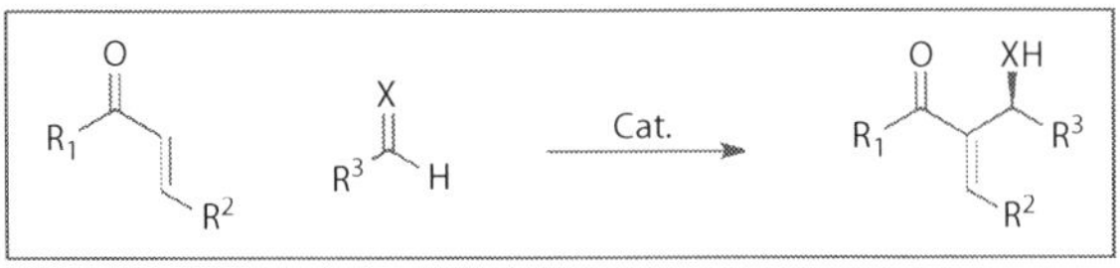

FIGURE 3.8 Amine-catalyzed Morita–Baylis–Hillman.

91). This reaction delivers products that contain the additional olefinic unit, which is a synthetically valuable handle that can be further used in a synthesis campaign and, in addition, can be driven to asymmetry. The overall transformation that is achieved is shown in Figure 3.8.

3.2.1.5 Phosphine Catalysis

Phosphine catalysis has been shown to be broadly applicable to a series of bond-forming events, which generally involves the anion of an electron-deficient π-system and a range of compatible nucleophiles (for selected reviews, see Refs. 92–94). The scope of electron-deficient π-system ranges from enals to ynals to allenes and, combined with a wide variety of suitable coupling partners, this catalysis mode provides convenient access to a relative breadth of product architectures. Many of the typical applications are formally related to the Morita–Baylis–Hillman reaction and, indeed, can provide an alternative catalytic framework for processes, which may be unreactive under amine catalysis. A summary of the typical reaction classes is provided in Figure 3.9.

3.2.1.6 Ketone/Iminium Ion-Catalyzed Epoxidation Reactions

Epoxides are highly useful synthetic intermediates that have found extensive use in organic chemistry and, consequently, their enantio-selective synthesis has inspired the development of a host of methodologies. A tranche of methods for the stereo-selective synthesis of epoxides have been generated in many of the branches of

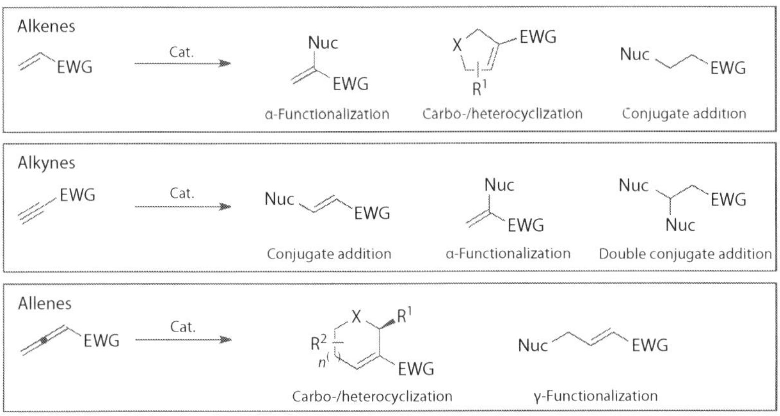

FIGURE 3.9 Illustrative summary of phosphine catalysis reaction classes.

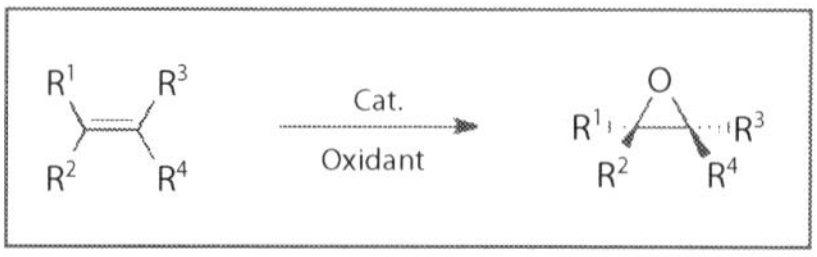

FIGURE 3.10 Ketone/iminium ion-catalyzed epoxidation.

organocatalysis. The use of Lewis acidic ketone and iminium ion catalysts provides novel and mechanistically distinct approaches to the enantio-selective epoxidation of a range of olefinic substrates (Figure 3.10) (for selected reviews, see Refs. 95 and 96).

3.2.2 Brønsted Acid and Brønsted Base Catalysts

3.2.2.1 Brønsted Acid Catalysis

Promotion of organic reactions by the action of Brønsted acids has been historically valuable and continues to enable some of the most important and widely used synthetic processes. Accordingly, the development of chiral Brønsted acid systems has provided a platform of enormous potential for asymmetric synthesis (for selected reviews, see Refs. 97–103). Catalysis of many of the classic acid-promoted reactions has become possible with high levels of asymmetric induction whereas new applications continue to provide access to novel chiral scaffolds with exquisite selectivity. A summary of typical reaction classes is provided in Figure 3.11.

3.2.2.2 Cinchona Alkaloid Catalysis

Cinchona alkaloids are one of the relatively few natural products that are used directly, without further functionalization, as catalysts in organocatalysis (others include amino acids used in aminocatalysis) (for selected reviews, see Refs. 104 and 105). The amine functionality of the parent quinuclidine system is both nucleophilic and basic and therefore enables catalysis of a diverse range of reaction classes involving nucleophile-driven mechanisms and deprotonative mechanisms including classic enolate-based chemistry (aldol, Mannich, etc.) and Morita–Baylis–Hillman-type reactions, respectively. In addition, the acid salts of cinchona alkaloids can be employed as catalysts for asymmetric protonation. Moreover, *N*-alkylated quaternary ammonium salt derivatives are excellent phase transfer catalysts (*vide infra*). A summary of typical reaction classes is illustrated in Figure 3.12.

3.2.2.3 Guanidine/Amidine Catalysis

Guanidines and amidines are relatively strong bases and their salts readily participate in accretive H-bonding interactions with putative substrates. Consequently, these catalysts are able to activate both nucleophilic and electrophilic species to enable enantio-selective versions of a variety of common and widely employed synthetic organic transformations, generally based on carbonyl containing compounds, such as aldol, Mannich, and conjugate addition processes (for selected reviews, see Refs. 106–109). A summary of typical reaction classes is provided in Figure 3.13.

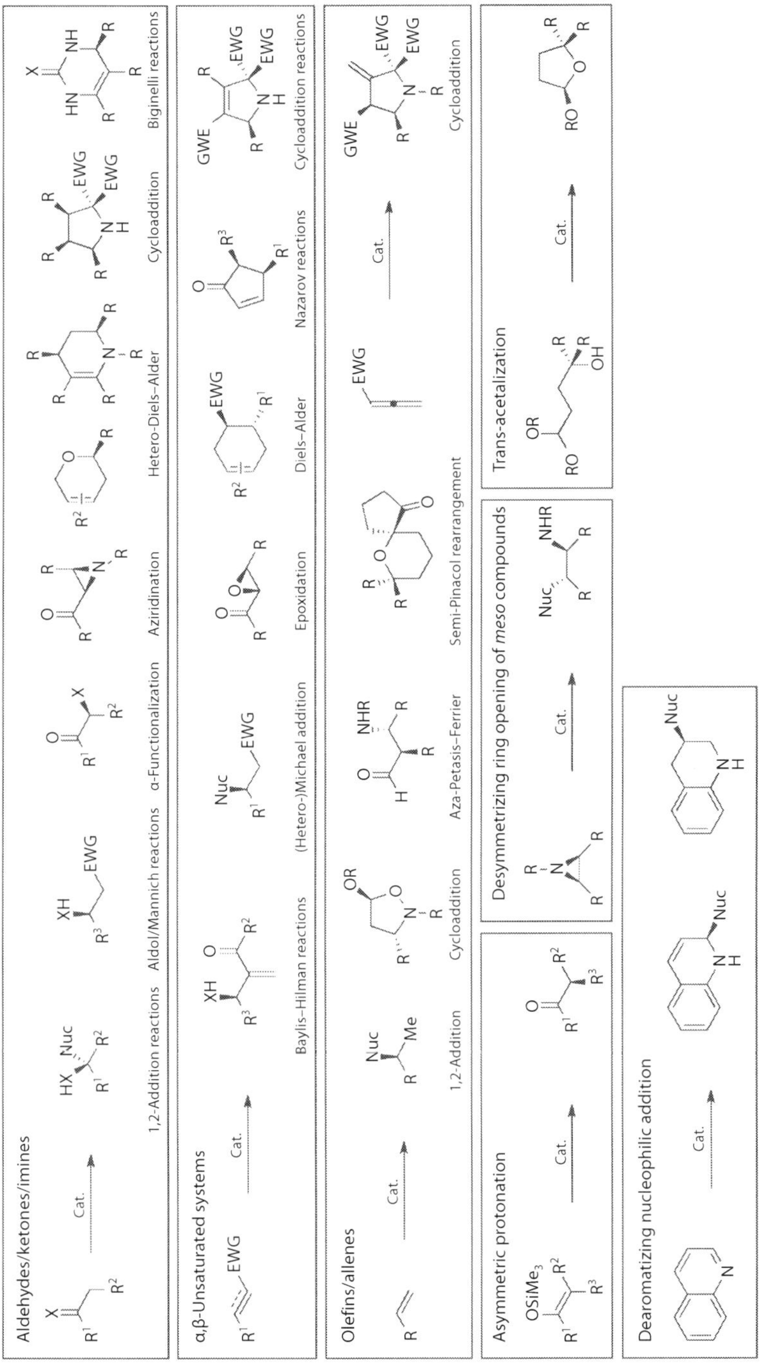

FIGURE 3.11 Illustrative summary of Brønsted acid catalysis reaction classes.

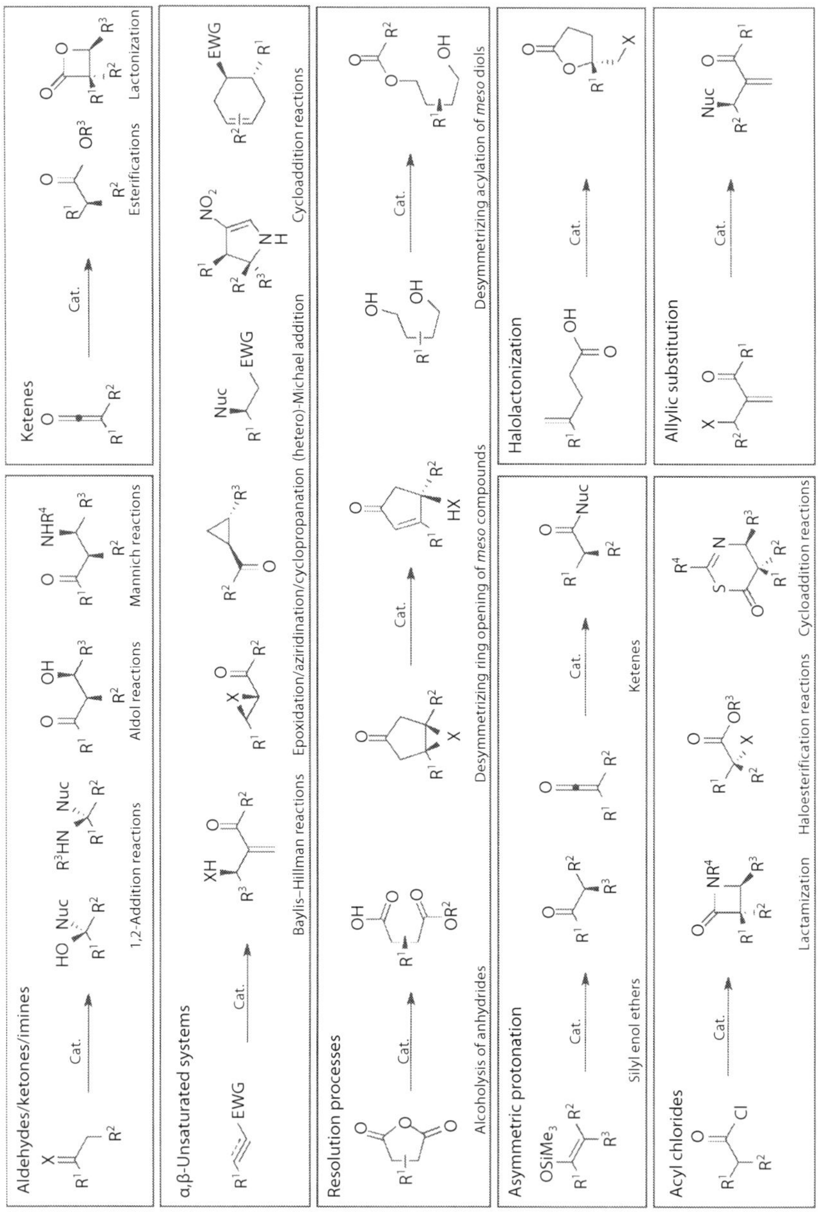

FIGURE 3.12 Illustrative summary of cinchona alkaloid catalysis reaction classes.

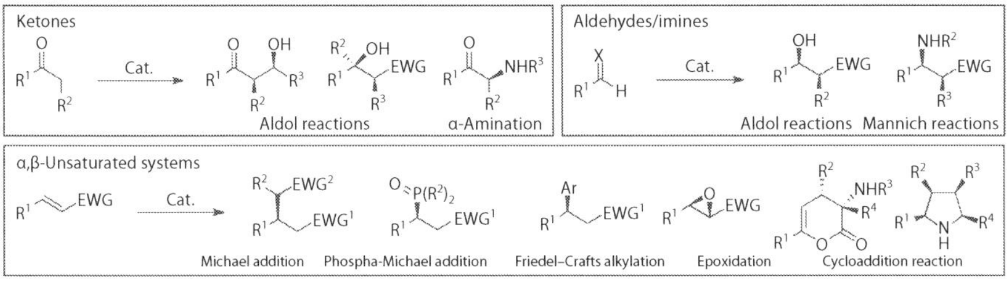

FIGURE 3.13 Illustrative summary of guanidine/amidine catalysis reaction classes.

3.2.2.4 H-Bonding Catalysis

The coordination of functional groups associated with high electron density through discrete H-bonds established through a suitable H-bond donor has provided a versatile platform for asymmetric catalysis. A variety of suitable H-bond donors have been shown to provide effective chiral frameworks for molecule activation and have facilitated the development of both asymmetric variants of well-established organic transformations as well as entirely new bond-forming processes[103] (for selected reviews, see Refs. 110–113). A summary of typical reaction classes in shown in Figure 3.14.

3.2.2.5 Bifunctional Catalysis

Judicious selection and combination of the attributes of two different catalyst activation modes provides the opportunity to develop bifunctional catalysts that can perform multiple roles to enable the desired bond-forming event (for selected reviews, see Refs. 114–117). In so doing, greater possibilities exist for the union of reacting components, providing more expedient access to new reactions and molecular architectures or to improve on the prevailing reactivity/selectivity profiles of existing individual catalysts. A summary of typical reaction classes is provided in Figure 3.15.

3.2.2.6 Phase Transfer Catalysis

Phase transfer catalysis enables reactions to be performed in a biphasic mixture (organic solvent/aqueous) and therefore offers the possibility of reducing the overall use of organic solvents in a given synthesis, which may assist in the development of an overall more sustainable process (*vide infra*) (for selected reviews, see Refs. 118–122). Generally used for transformations associated with acidic functionalities, such as activated methylene compounds, typical phase transfer catalysis reactions include asymmetric versions of the classic carbon–carbon and carbon–heteroatom bond-forming reactions of carbonyl chemistry. A summary of the representative reaction classes is shown in Figure 3.16.

3.3 ANALYSIS OF REACTION VARIABLES OF ORGANOCATALYTIC PROCESSES

Analysis of the variables (reagents, solvents, stoichiometry, temperature, time, etc.) of a particular transformation can provide insight into how well aligned the reaction is with modern concepts of sustainable chemistry[28–31] (for useful information, see

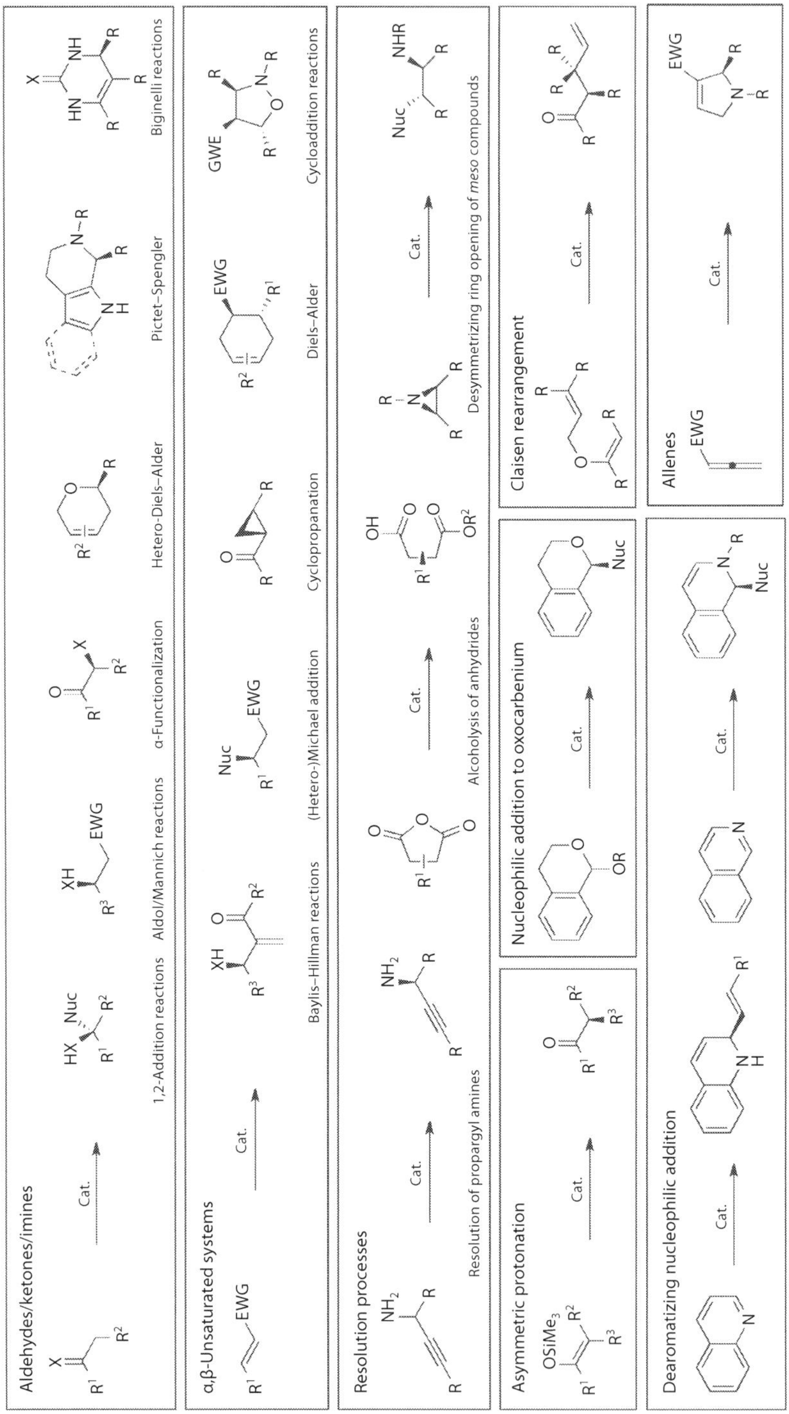

FIGURE 3.14 Illustrative summary of H-bonding catalysis reaction classes.

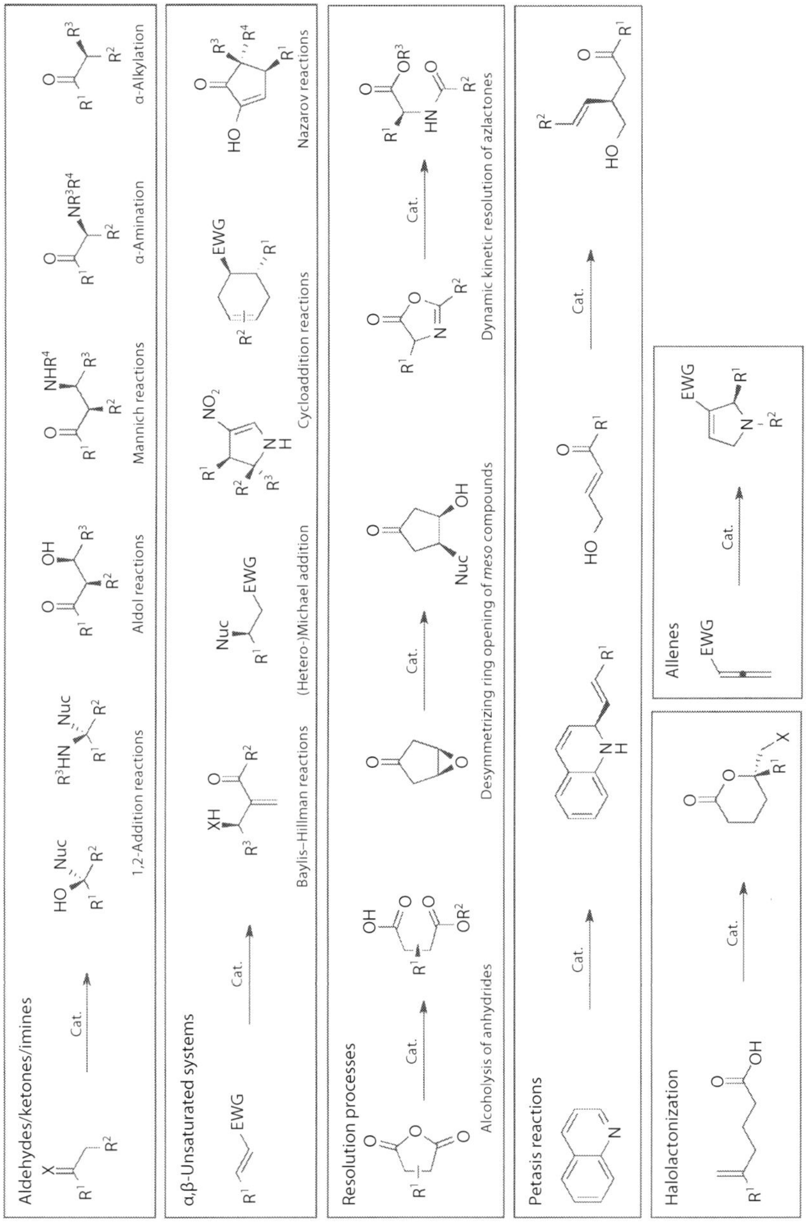

FIGURE 3.15 Illustrative summary of bifunctional catalysis reaction classes.

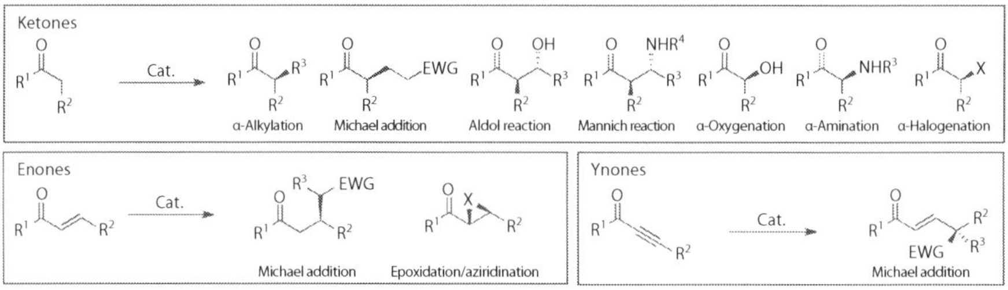

FIGURE 3.16 Illustrative summary of phase transfer catalysis reaction classes.

Refs. 123–145). In addition, analysis of a larger dataset of representative reactions can provide a more global insight into the compatibility of the wider class or field with the ethos of green chemistry. Consideration of the requisite variables in a given organocatalytic reaction, such as driving down the quantity of reagents used or avoiding solvents with serious safety, health, and environment (SHE) issues, prolonged reaction times, etc., can assist in providing a more sustainable overall process. The following sections discuss several of the main variables of organocatalytic reactions in the context of how well they dovetail with the principles of sustainable chemistry.

3.3.1 Stoichiometry of Reagents

The first of the Twelve Principles of Green Chemistry is waste prevention.[32] Balancing the stoichiometry of reagents used in a reaction is one of the simplest methods of improving the overall efficiency of a given process through lowering the waste output. The ideal scenario would be where equivalent molar ratios of reagents react, in the absence of additives, to provide 100% yield of the product (other considerations such as energy, solvent, isolation, etc., aside). As stoichiometries become increasingly unbalanced, or where excessive quantities of additives are used, the quantity of waste associated with a particular process increases commensurately. This negatively affects the sustainability of the process, which could have a commensurate effect on the uptake or adoption.

In the context of organocatalysis, broadly speaking, waste generation arising from excesses of a particular reagent or, indeed, reagents is typically high. Many reactions are exquisitely selective and provide high yields of product. However, the necessary components of the reaction scheme are frequently highly unbalanced and often rely on the presence of large amounts of ancillary additives or promoters. For example, within the enamine class of aminocatalysis, it is not uncommon for reactions involving nucleophilic species derived from activated methylene compounds (for example, ketones and nitoalkanes) to be conducted using the nucleophilic donor as the reaction solvent. Similarly, excesses of the corresponding electrophilic components are also frequently employed to achieve high levels of conversion/selectivity. Moreover, many of these electrophilic or nucleophilic components exhibit significant SHE issues and can require care in preparation (if necessary), handling, storage, use, and disposal.

Analogously, many reactions involve large excesses of organic and inorganic acids, bases, oxidants, and reductants (it should be noted that there are overarching sustainability issues with many of these). For some reactions, these supplementary reagents are not necessarily benign. For example, *N,N*-diisopropylethylamine (DIPEA, Hunig's base) and K_2CO_3 are commonly used Brønsted basic reagents within a variety of organocatalytic processes, such as cinchona alkaloid catalysis and phase transfer catalysis, respectively. At a cursory level, these are comparatively less problematic than when compared with reactive bases such as KH, which has been used for catalyst generation in carbene catalysis from the requisite azolium precursor systems.

Overall, waste generation, as a consequence of stoichiometry within organocatalysis in the general sense, is currently a necessary challenge to overcome as we progress toward a more sustainable technology. The development of more efficient processes based on a more balanced stoichiometry does present issues, the largest of which is, perhaps, cultural; it is unfortunately not often considered the most academically interesting pursuit and is therefore more likely to be addressed only after an organocatalytic reaction becomes integral to some part of a large-scale industrial synthesis campaign where such considerations are more integral to developing a more economical process.

3.3.2 Solvent Use

The use of an organic solvent or cosolvent is typically essential for a successful outcome from most organocatalytic reactions. Solvent is one of the primary contributors to the waste output from the majority of chemical processes.[146–153] Indeed, in the context of pharmaceutical manufacture, solvent represents the largest fraction of the total material input and is one of the principal contributors to the waste output of overall synthetic operations. It is therefore not surprising that considerable efforts continue to be made to lower the effect of solvents through recycling, reduction, or replacement.

Recycling solvent through recovery (e.g., through distillation) will have associated necessary inputs (energy and materials) and can only be determined to be viable following a thorough life cycle analysis (LCA).[127] Recycling then will only be beneficial after establishing the net efficiency benefits versus disposal. As previously discussed, the majority of organocatalytic reactions are developed in the academic environment and on a small (mmol/sub-mmol) scale. Consequently, solvent recovery and reuse tends not to be a primary concern and will only likely be part of a larger institutional recycling policy, if such a policy is in place. Similarly, solvent recycling is not routine in discovery phase industrial chemistry—such considerations are typically only found in larger-scale processes and then only if LCA suggests a net benefit.

Reduction in volume relies on operating reactions at increased concentration or more effective use of solvents during workup, purification, and isolation. Efficiencies in the isolation process may be possible and is certainly an option that can be explored. However, increasing reaction concentration is less facile and will likely require some level of additional optimization—many organocatalytic processes have been extensively optimized and deviation from the source conditions may profoundly affect the conversion/selectivity.

Replacement of hazardous solvents to achieve greater sustainability or to reduce both environmental and operational costs is perhaps a more straightforward approach and, indeed, has become a key emerging consideration within the chemical industry.[146–153] A number of reports have emerged from industrial companies detailing the drivers and requirements for replacement of conventional, less sustainable solvents with those that are viewed, for various reasons, more benign.

Although organocatalysis has matured simultaneously with green chemistry, the uptake and employment of greener solvents in this area has been relatively slow. Organocatalytic reactions are performed in a huge variety of organic solvents, many of which are entirely incompatible with sustainable chemistry principles. This can be attributed, at least in part, to the emergent nature of the field: as described above, the field of organocatalysis has been the subject of an extraordinary rate of expansion. Organocatalysis is currently most frequently practiced in academic laboratories (although it should be noted that industrial uptake of various techniques is increasing) where the use of alternative solvents is less of a priority. Having said this, a number of studies have used more sustainable solvents [e.g., acetone, cyclopentyl methyl ether (CPME), EtOAc, 2-methyltetrahydrofuran (2-MeTHF), and *i*-PrOH] from the outset and others have attempted to improve the green credentials of existing processes through replacing less desirable solvents with those that are more sustainable (including reactions in/on water).[154] A survey of representative solvent use within the branches of organocatalysis is given in Table 3.1.

From the representative solvent distribution above, it is clear that solvent use in organocatalysis is far from consistent with green chemistry principles. The overwhelming majority of reactions are conducted in solvents, which have either serious (chlorinated, average 27% of reactions surveyed) or moderate (PhMe, 28% of reactions surveyed) SHE concerns. Although the distribution of solvents varies slightly from branch to branch, the data reveals that the most common solvents employed for organocatalytic reactions are those that are generally targeted for replacement from a sustainability perspective.

To assist in replacing solvents, equielutropic series have been constructed; however, these tend to be heavily skewed toward solvents with major regulatory or toxicological issues, for example chlorinated, toluene, and hexane,[155] which may be influential given the preponderance of these systems in organocatalytic processes. Indeed, efforts have been made by investigators in the organocatalytic arena to employ solvents that are more in keeping with green chemistry. These currently number in the small minority. It should be noted that this analysis has only taken the reaction solvents into account—an additional consideration is the solvents used as part of the isolation process. As reports continue to emerge and provide more effective guidance in the solvent replacement area, the uptake of more benign solvents in organocatalysis may increase.

3.3.3 Catalyst Loading and Selection

3.3.3.1 Catalyst Loading

The performance of a particular organocatalytic reaction can often be directly related to the loading of the catalyst. In many cases, loadings of organocatalysts can

TABLE 3.1
Representative Solvent Use in Organocatalytic Activation Modes

Branch	Chlorinated (%)	1,4-Dioxane (%)	DMF (%)	DMSO (%)	H_2O (%)	MeCN (%)	THF (%)	PhMe (%)	Other (%)	*n*
Aminocatalysis	24	4	8	10	11	6	10	10	27	1713
Acyl transfer	49	3	–	–	–	–	2	33	15	187
Carbene catalysis	21	–	–	4	4	2	31	36	2	352
Amine-catalyzed Morita–Baylis–Hillman reactions	10	–	20	2	3	13	13	20	19	240
Phosphine catalysis	25	1	–	–	>1	4	28	21	20	460
Ketone/iminium-catalyzed epoxidation	9	4	–	–	5	34	–	–	48[a]	247
Brønsted acid catalysis	36	2	–	–	–	1	2	36	23	898
Cinchona alkaloid catalysis	37	4	6	–	–	4	4	18	27[b]	279
Guanidine/amidine catalysis	16	–	–	–	12	3	13	33	23	329
H-Bonding catalysis	33	–	>1	–	1	3	6	34	23	785
Bifunctional catalysis	35	>1	1	–	–	4	7	41	12	504
Phase transfer catalysis	27	2	–	–	[c]	–	–	55	16	535
Average	27	2	3	1	3	6	10	28	20	544

Note: From analysis of data presented in Reference 26.

[a] Of these, 32% were conducted in dimethoxymethane and 16% in DME.

[b] Of these, 17% were conducted in Et_2O.

[c] All reactions were biphasic; water was used in all examples.

be relatively high by comparison to metal-catalyzed processes. Indeed, 20 mol% of organocatalyst is not uncommon and is often the starting point for many practitioners when developing a new process. Organocatalysts can be more favorable than existing metal-based catalysts because the tolerance for organic impurities in end products, such as pharmaceuticals, is higher than that of many metal-based impurities, which often require quantification to the parts per billion (ppb) level. However, lowering organocatalyst loading remains a priority as (i) the catalysts must be prepared or purchased and (ii) the catalysts must still be separated from the product(s) to a specific level and, LCA permitting, recycled. In addition, the absolute toxicity profiles of the majority of organocatalysts remain unknown.

Although the tolerance for impurities of organic catalysts may be higher in end products, the removal and reuse of these catalysts may not be entirely straightforward. On a small scale, such as within academia or a discovery phase industrial setting, chromatographic separation is routine and the catalyst can be removed while the product is purified. However, on a large scale such as within process chemistry or primary manufacture, chromatography is less routine. Metal-based catalysts may simply be removed by careful filtration while retaining the product in solution. However, organocatalysts will likely be removed through crystallization methods, either by precipitation of the product or the catalyst from the crude reaction milieu. The likelihood of success of this separation will ultimately depend on the nature of the catalyst.

3.3.3.2 Catalyst Selection

The selection of an appropriate catalyst is crucial to the successful outcome of a reaction. There are aspects of many organocatalytic transformations that require a particular catalyst type be employed. For example, many proline-based reactions will only operate with proline due to the proximity of the carboxylic acid motif. Similarly, SOMO processes are exclusively effective with imidazolidinone catalysts, etc. Additionally, where relevant, the acid cocatalyst can have a profound effect on the reaction outcome and often the same amine catalyst will be used as a different acid salt for different reactions.

Amino acid and cinchona alkaloid catalysts are natural compounds, which are readily available. A selection of some of the more broadly applicable catalysts for each of the activation modes have become commercially available (e.g., selected imidazolidinones and NHC precursors); however, when required, more specialized variants must be prepared. The syntheses of organocatalysts can range from a few relatively trivial steps, such as in the case of many imidazolidinone catalysts or NHCs, to a substantial number of steps, including chiral resolution or separation, which may be challenging to perform on scale. Indeed, catalyst synthesis may require more synthetic steps than for the entire target molecule synthesis in which the catalyst is planned for use. Accordingly, although organocatalysis presents methods for achieving stereocontrolled bond formations that may otherwise be less accessible through other methods, one must carefully consider any necessary catalyst synthesis and attendant sustainability issues as part of the synthetic strategy.

Perhaps one of the key benefits as compared with conventional (metal) catalysis is that organocatalysis can generally be defined as truly catalytic in nature: the majority of organocatalysts may be recovered unchanged at the end of the reaction and reused if desired. This is in contrast to the majority of metal-based catalysts, which are often precatalysts that cannot be recovered in their original state (oxidation, ligation, etc.). Although it may be possible to recover metal residues from a reaction and convert these back to the original precatalyst, this will require further processing. However, even though most organocatalysts will not require significant processing before reuse, LCA would be necessary to ascertain the benefits of any proposed recycling.

Focusing on specific exemplars, there are a number of privileged catalyst scaffolds in each branch; however, the diversity of analogues is very broad indeed. An illustrative selection of typical catalyst architectures for each of the branches of organocatalysis is provided in Charts 3.1 and 3.2.

3.3.4 Reaction Temperature and Time: Energy Efficiency

Minimizing the energy consumption of a particular reaction to reduce the associated economical and environmental impact represents the sixth green chemistry principle.[32] Indeed, any thorough LCA must address the effect of energy on the overall efficiency of a transformation because energy is typically derived from nonrenewable resources, with attendant waste production. Studies have attempted to assess the consumption of energy in reactions that employ different techniques (conventional stirring/heating, microwave, etc.) to provide guidance as to the most energy-efficient methods.[129] Clearly, the energy use of a chemical reaction can be lowered by operating at ambient temperature and pressure, and as swiftly as is practicable to avoid long residency times in the reaction vessel. Of course, the reaction itself is only part of the energy consumption: work-up, isolation, and analysis procedures will typically require additional energy input, particularly when ancillary instrumentation is taken into consideration.

Organocatalytic reactions are typically conducted at one atmosphere and therefore generally avoid any energy wastage associated with increased pressures. However, it is evident from the primary literature that a broad range of reaction temperatures and times are employed.

The use of higher or lower temperatures is not simply specific to a particular branch or type of transformation but can be highly reaction-specific. Consequently, some reactions can be carried out conveniently at room temperature whereas many are cooled (0°C and below) and relatively few are heated above ambient temperature. The routine use of lower temperatures can be attributed to the nature of asymmetric catalysis and the associated need to reconcile selectivity and reactivity. Improved energy efficiency may be achieved at temperatures closer to ambient; however, the enantio-selectivity or diastereoselectivity may be compromised. Product purification will then require additional manipulations (separation by HPLC, derivatization, fractional crystallization, etc.) necessitating additional inputs (solvents and reagents) and processing, which will negatively affect the overall efficiency of the reaction. Consequently, lower energy efficiency in favor of increased reaction efficiency may be more straightforward and practicable. Typical temperature ranges employed for the various branches of organocatalysis are shown in Table 3.2.

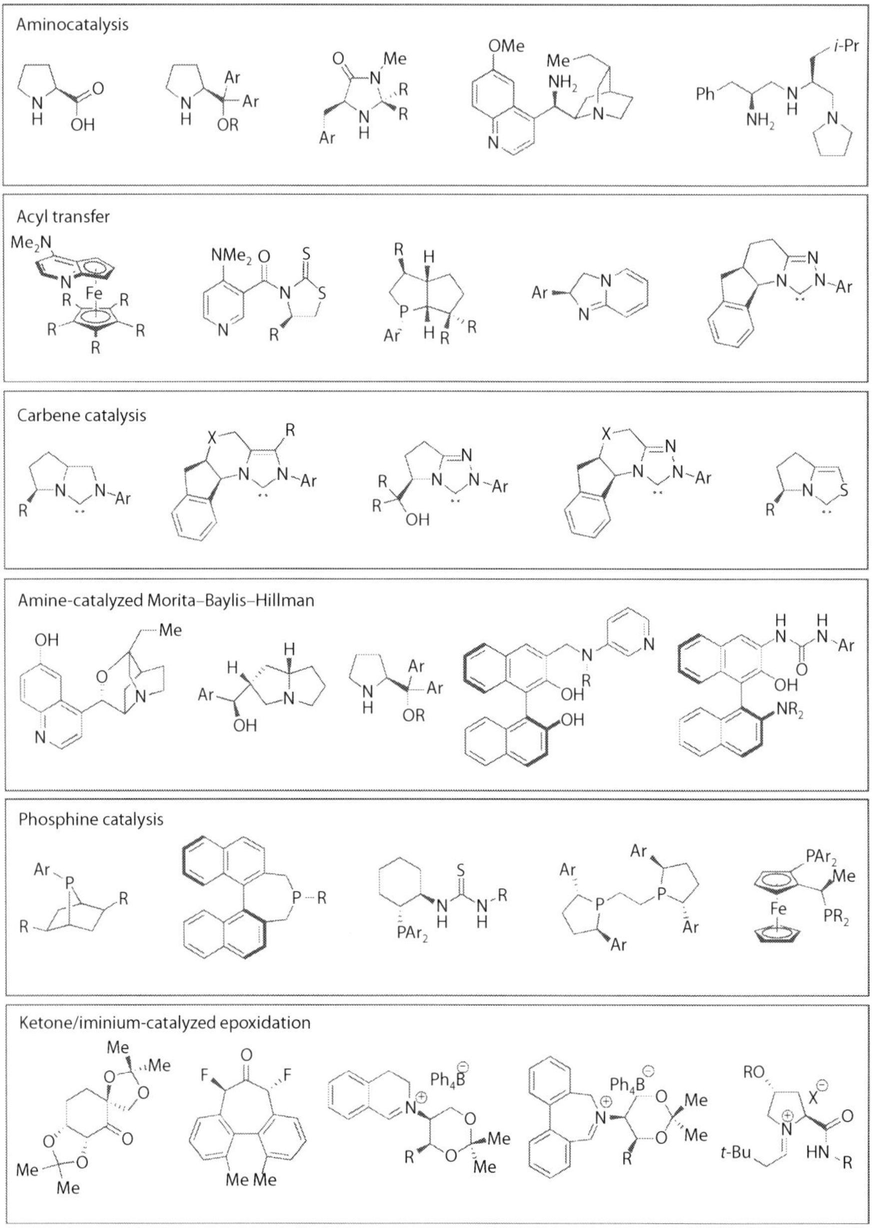

CHART 3.1 Examples of typical Lewis acid and Lewis base catalysts.

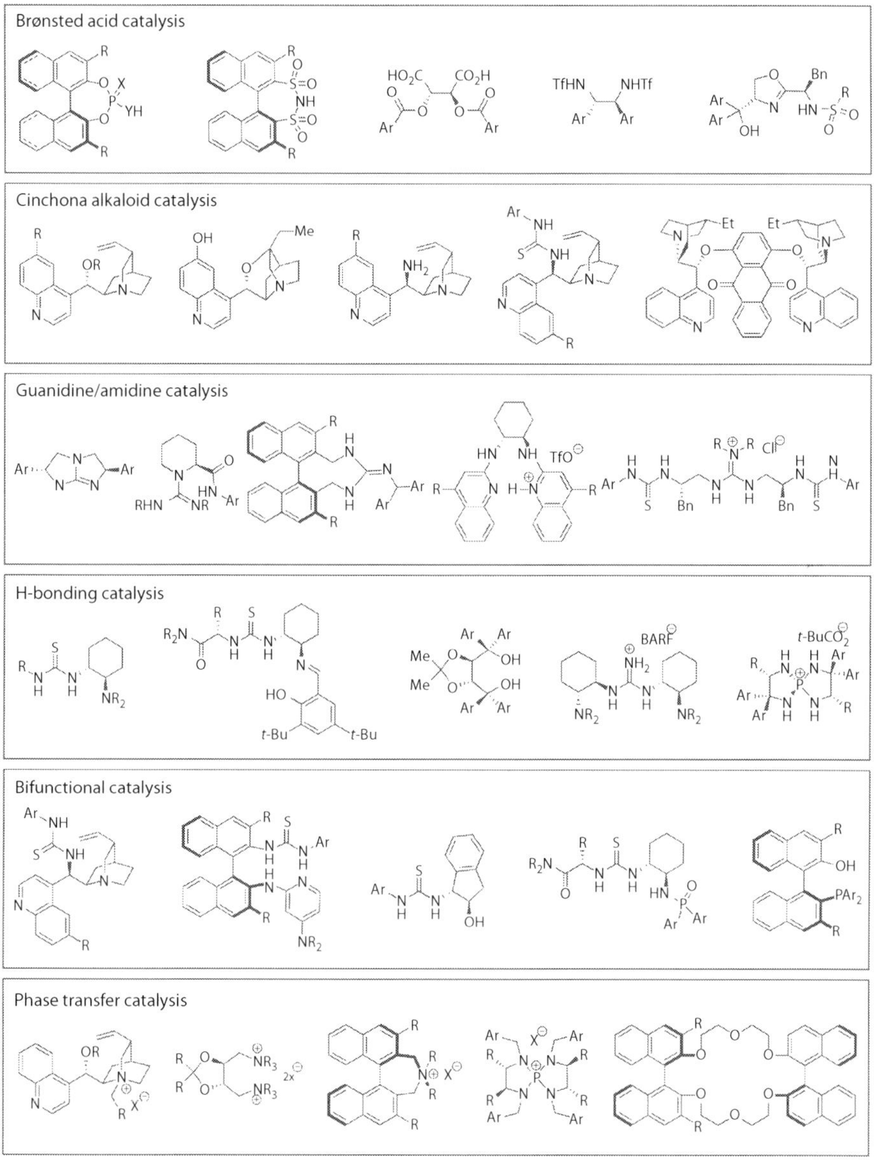

CHART 3.2 Examples of typical Brønsted acid and Brønsted base catalysts.

Similarly, a broad range of reaction times is used. Many organocatalyzed reactions are fast and are completed in minutes whereas some require days to reach completion; of course, this is highly dependent on the reaction temperature. Accordingly, energy efficiency will vary directly with the residency time. Clearly, longer residency times coupled with temperatures above/below ambient will be less favorable

TABLE 3.2
Representative Temperature Use in Branches of Organocatalysis

Organocatalysis Branch	Temperature Range	Modal Temperature	Average Temperature	*n*
Aminocatalysis	−83°C to +170°C	RT	Approx. 7°C	1713
Acyl transfer	−78°C to 75°C	0°C	Approx. −17°C	187
Carbene catalysis	−78°C to 110°C	RT	Approx. 23°C	352
Amine-catalyzed Morita–Baylis–Hillman Reactions	−55°C to 25°C	−30°C	Approx. −13°C	240
Phosphine catalysis	−60°C to 140°C	RT	Approx. 17°C	460
Ketone/iminium-catalyzed epoxidation	−78°C to RT	0°C	Approx. 4°C	247
Brønsted acid catalysis	−88°C to 130°C	RT	Approx. 4°C	898
Cinchona alkaloid catalysis	−78°C to 80°C	RT	Approx. −7°C	279
Guanidine/amidine catalysis	−80°C to 40°C	−20°C	Approx. −22°C	329
H-Bonding catalysis	−98°C to 50°C	RT	Approx. −19°C	785
Bifunctional catalysis	−80°C to 80°C	RT	Approx. −6°C	504
Phase transfer catalysis	−80°C to 80°C	0°C	Approx. −13°C	535

Note: From analysis of data presented in Reference 26.

from an energy consumption standpoint. Similar to reaction temperature, decreasing what may be an unfavorable reaction time may have a beneficial effect on energy efficiency but may lead to decreased reaction efficiency, leading to the associated downstream effects described in previous sections.

A cursory examination of the data in Table 3.2 indicates that there is a high degree of probability that organocatalytic processes can embrace the sixth principle of green chemistry.[32] For many of the reaction manifolds, the modal temperature of operation is ambient. Evidently, certain reaction classes operate below this with commensurate effect on reaction time. Consequently, the energy efficiency of organocatalytic reactions is variable and is heavily dependent on both this branch of catalysis as well as on the individual reaction subtype. In a general sense, the majority of organocatalytic reactions need to be cooled and reaction times can be lengthy. Consequently, from an energy conservation perspective, further work is required to improve efficiencies here. However, and as stated above, the energy efficiency is only one part of a thorough LCA and this must be placed in context with alternative methods for achieving the same bond construction, which may be more lengthy and therefore have a poorer overall efficiency.

3.4 SUSTAINABILITY METRICS AND ORGANOCATALYSIS

The overall sustainability of individual chemical transformations and overall processes can, to a certain extent, be quantified through the use of a suitable evaluation tool. This is typically achieved through calculation of one or a combination of several metrics, which provide a numerical output that can then be measured against either preceding versions of the same technology or contemporary methods for achieving the same bond formation. A range of metrics have been designed and developed, each with specific outputs that inform on a range of aspects of a particular reaction or process.[28–31,123–144] The most comprehensive method is LCA, which has several variants.[126] LCA is a "cradle to grave" analysis of a particular product that evaluates all associated inputs and outputs and is routinely employed for large-scale industrial processes. However, as discussed above, the majority of organocatalytic reactions reported in the literature are performed in the academic environment on comparatively a much smaller scale and where the information required to perform such an analysis is not frequently captured (for example, the quantities of solvent used during purification). Having said this, a series of additional, less comprehensive tools have been developed and can provide insight into and assist in the evaluation of most chemical reactions. These metrics require a range of different inputs and an assessment can be performed using the available data, for example, from the supporting information, which accompanies a primary report of a reaction of interest.

There are several metrics that are both commonly used and that are suitably flexible to enable application to transformations without all the data required for a full LCA. In particular, atom economy (AE),[123] E Factor,[124,132] reaction mass efficiency (RME),[126] and molar efficiency (Mol. E)[145] can be readily used to interrogate the efficiency and environmental impact of smaller scale (academic) chemical processes in lieu of a full LCA.

AE serves to determine the efficiency of a chemical reaction with regard to how many atoms from the starting materials reside within the product, with 100% being ideal. However, AE disregards the chemical yield of the reaction and any associated wastage. By comparison, RME relates the mass of components employed in a reaction to the mass of the desired product output and, accordingly, gives more detail over the associated waste. Similar to AE, the ideal reaction will have a value of 100%. One limitation is that RME only factors the stoichiometric reagents and disregards any substoichiometric additives or catalytic agents. E Factor is by far the more comprehensive analysis and compares the mass of product to the mass of waste produced for a given process, taking into account all reaction components and the associated waste, including any associated purification steps, with the ideal process registering a value of zero. E Factor calculations can be usefully modified to use fewer parameters (e.g., based solely on the reaction equation) if the information for a full analysis is unavailable, offering a restricted E Factor, which can be used to evaluate a process of interest. Mol. E has recently been introduced to overcome the possible limitations of mass-based efficiency matrices. Because Mol. E is based on molar values for all components (stoichiometric and substoichiometric) of a reaction, the calculated values are readily compared across different scales or variants of one particular process and, importantly, across different reaction classes. Mol. E values give a measure of both the efficiency of the process and also the waste output and is, consequently, more holistic than AE and RME.

In the context of organocatalysis, and as outlined above, being primarily developed in academia, the captured experimental data is typically insufficient to enable calculation of the more extensive metrics (e.g., LCA or full E Factor) for any of the

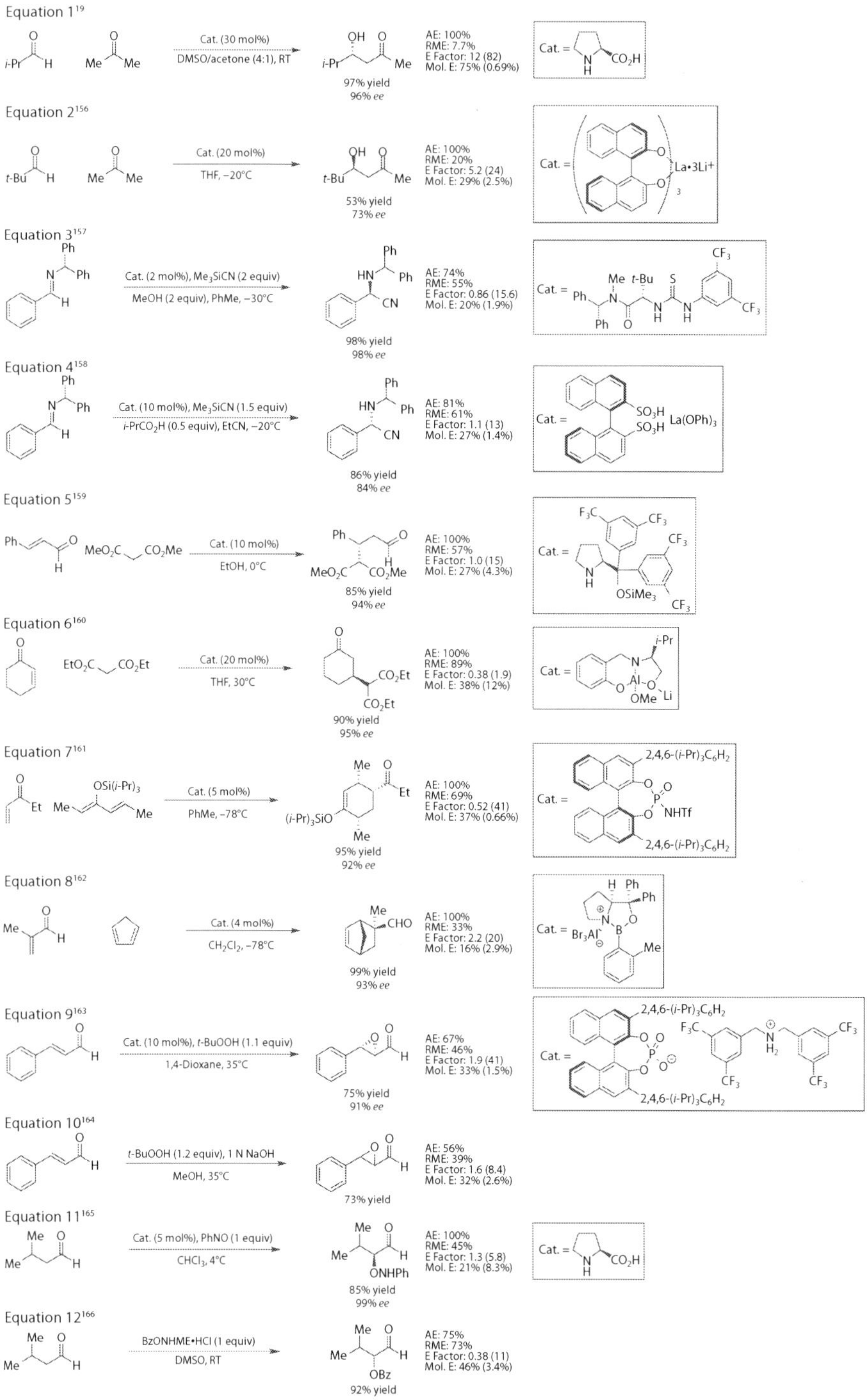

CHART 3.3 Comparison of organocatalytic versus conventional bond-forming processes.

selected transformations. However, AE, RME, Mol. E, and a restricted E Factor (based on the catalytic and stoichiometric components) can be calculated based on the available data from the experimental procedures and supporting information associated with the parent publications.

To place organocatalytic procedures in context with contemporary methods in the field of organic synthesis, Chart 3.3 displays several examples of organocatalytic reactions alongside conventional methods for achieving the same or similar transformations with AE, RME, restricted E Factor, and Mol. E metrics calculated for each. These examples are paired based on an organocatalytic procedure and a conventional procedure for the same or equivalent transformation (i.e., 1 should be considered with 2, 3 with 4, etc.): odd-numbered equations represent organocatalytic processes whereas the even-numbered equations are the conventional methods. The values in parentheses for restricted E Factor and Mol. E indicated the value when solvent is included in the calculation.

Although not exhaustive of the variety of procedures available for each transformation, from the values for the metrics calculated for these examples, it can be seen that the organocatalytic methods generally tend to perform similarly to conventional methods. Values for AE, RME, restricted E Factor, and Mol. E are broadly comparable within each reaction class with no prominent examples of a preferable method. As discussed in previous sections, the solvent selection in organocatalysis is not necessarily competitive with conventional procedures and, from an energy consumption perspective, temperatures can be similarly variable as compared with conventional methods. Selectivities and chemical efficiencies are also broadly comparable. Based on this, from an efficiency metrics perspective, organocatalytic processes perform similarly to the more conventional approaches.

Having stated this, and as can be seen from the majority of contemporary reactions analyzed in Chart 3.3, a primary benefit of organocatalysis is the reduced reliance on metallic additives/catalysts, leading to lower levels of metallic waste and, indeed, no need to employ elements that are perhaps of particular concern due to scarcity.[167] For example, lanthanum (Equations 3.2 and 3.4; Chart 3.3) is a metal commonly used within organic synthesis but has significant future supply risks. A second primary benefit that is overlooked from the above analysis is the opportunity that organocatalysis provides for direct and selective bond formations that were hitherto unknown, thereby improving overall efficiency and diversity in terms of molecule synthesis. The previous examples focus on specific transformations with conventional alternatives. However, there are many organocatalytic reactions that do not have a comparable conventional alternative and accordingly the appropriate comparisons cannot be made. In the context of a complete synthesis of a target molecule, these reactions could facilitate more expedient access to the desired compounds with a concomitant effect on waste output.

3.5 SUMMARY AND OUTLOOK

Since its inception in 2000, the domain of organocatalysis has matured into a vast field that encompasses numerous activation modes each with a multitude of specific bond-forming reaction types. Currently, organocatalysis is primarily developed and practiced most extensively in the academic arena and is performed on a relatively small scale. Although several advantages are apparent, efficiency indices of the processes examined here are typically comparable with conventional methods. However, it should be borne in mind that many organocatalytic processes do not have a conventional counterpart. Perhaps one of the most informative indicators of the current level of sustainability of organocatalysis is the uptake of organocatalytic activation on the large scale, that is, within process chemistry operations. Early stage drug discovery and agrochemistry are typically less concerned with the sustainability credentials of the chemistry being employed; however, it should be noted that sustainability is becoming increasingly important in the discovery phase.[168] Encouragingly, organocatalytic procedures are continuing to advance from these small-scale applications into practical methods for the synthesis of key commodities, such as pharmaceuticals. It is at this production scale where the sustainability aspects of organocatalytic technology can be most effectively gauged. As discussed previously, small-scale academic or discovery phase studies are of limited use here since, by their nature, these tend to be intrinsically inefficient from a sustainability perspective: the principal goals typically focus on increased selectivity and yield—the global effect of all necessary inputs and the potential for recovery/recycle, and so on, are generally not frequently considered *a priori*. However, on preparative scales, such as within process chemistry, global efficiency is paramount and optimization of selectivity and yield is coupled with sustainability from the outset to manage resources and waste output.

The majority of academic studies in the area of organocatalysis is knowledge-driven and primarily focus on the development of new bond formations and currently has somewhat limited attention directed toward producing reaction manifolds that are aligned with the principles of sustainability. Improving the sustainability credentials of existing organocatalytic transformations is typically viewed as less interesting; however, this is essential for the translation of these reactions into efficient processes that can be deployed on a production scale. It is reassuring and highly positive for the advancement of the area that studies continue to emerge, which have this sustainability agenda at their core.[169]

As identified above, there are several key areas that require additional research efforts to improve the overall sustainability of organocatalysis in a general sense:

i. Solvent selection. The majority of current solvents employed are typically inconsistent with green principles. Opportunities for research can be found in effective alternative solvents for existing reactions and embedding greener solvents at the outset of new methodology development.
ii. Catalyst loading. Although there are numerous examples of low loadings of organocatalysts, the majority of catalyst loadings are very high indeed, particularly in comparison with metal-catalyzed processes. The tolerance of organic impurities within end products, such as pharmaceuticals, is higher than for many metal species; however, the separation of organic compounds on a large scale may be problematic, depending on the catalyst and other reaction components. Many organocatalysts also require significant synthetic efforts in their own preparation and this will require analysis from a

sustainability perspective. Driving down the catalyst loading will assist in alleviating sustainability pressures arising from this essential component.

iii. Relating to (ii), the organocatalysts are responsible for the selectivity of a particular transformation. Improved catalysts may provide increased selectivity at temperatures closer to ambient and, therefore, offer improvements in terms of energy consumption.

iv. Unfavorable stoichiometries are often used to ensure good reaction performance. However, this leads to problems with waste output. Increased efforts are required to reconcile the use of super-stoichiometric reagents with sustainability.

Perhaps one of the simplest contributions researchers can make in this overall area and which will affect all areas (i–iv) is the provision of all data associated with a particular study. Often only the best results are disseminated in the literature and, consequently, it is likely that a vast amount of helpful data exists that has not been published. If this data were made available, it would prevent the repetition of experiments that have already been performed and will facilitate progress in this area in a general sense.

Overall, ensuring sustainability is a consideration from the outset of a particular methodological development study that will increase engagement and encourage further debate over the efficiency of the staple processes of synthetic chemistry in this widely practiced area.

REFERENCES

1. Ahrendt, K. A.; Borths, C. J.; MacMillan, D. W. C. *J. Am. Chem. Soc.* 2000, *122*, 4243–4244.
2. Komnenos, T. *Justus Liebigs Ann. Chem.* 1883, *218*, 145–167.
3. Knoevenagel, E. *Chem. Ber.* 1894, *27*, 2345–2346.
4. Langenbeck, W.; Sauerbier, R. *Chem. Ber.* 1937, *70*, 1540–1541.
5. Stork, G.; Terrell, R.; Szuszkovicz, J. *J. Am. Chem. Soc.* 1954, *76*, 2029–2030.
6. Hajos, Z. G.; Parrish, D. R. DE2102623, 1971.
7. Eder, U.; Sauer, G.; Wiechert, R. *Angew. Chem. Int. Ed. Engl.* 1971, *10*, 496–497.
8. Woodward, R. B.; Logusch, E.; Nambiar, K. P.; Sakan, K.; Ward, D. E.; Au-Yeung, B.-W.; Balaram, P. et al. *J. Am. Chem. Soc.* 1981, *103*, 3210–3213.
9. Woodward, R. B.; Logusch, E.; Nambiar, K. P.; Sakan, K.; Ward, D. E.; Au-Yeung, B.-W.; Balaram, P. et al. *J. Am. Chem. Soc.* 1981, *103*, 3213–3215.
10. Woodward, R. B.; Logusch, E.; Nambiar, K. P.; Sakan, K.; Ward, D. E.; Au-Yeung, B.-W.; Balaram, P. et al. *J. Am. Chem. Soc.* 1981, *103*, 3215–3217.
11. Yamaguchi, M.; Yokota, N.; Minami, T. *J. Chem. Soc., Chem. Commun.* 1991, 1088–1089.
12. Yamaguchi, M.; Shiraishi, T.; Hirama, M. *Angew. Chem. Int. Ed. Engl.* 1993, *32*, 1176–1178.
13. Yamaguchi, M.; Shiraishi, T.; Igarashi, Y.; Hirama, M. *Tetrahedron Lett.* 1994, *35*, 8233–8236.
14. Kawara, A.; Taguchi, T. *Tetrahedron Lett.* 1994, *35*, 8805–8808.
15. Yamaguchi, M.; Shiraishi, T.; Hirama, M. *J. Org. Chem.* 1996, *61*, 3520–3530.
16. Yamaguchi, M.; Igarashi, Y.; Reddy, R. S.; Shiraishi, T.; Hirama, M. *Tetrahedron* 1997, *53*, 11223–11236.

17. Wang, Z.-X.; Tu, Y.; Frohn, M.; Zhang, J.-R.; Shi, Y. *J. Am. Chem. Soc.* 1997, *119*, 11224–11235.
18. Corey, E. J.; Grogan, M. J. *Org. Lett.* 1999, *1*, 157–160.
19. List, B.; Lerner, R. A.; Barbas, C. F., III *J. Am. Chem. Soc.* 2000, *122*, 2395–2396.
20. Berkessel, A.; Gröger, H. *Asymmetic Organocatalysis: From Biomimetic Concepts to Applications in Asymmetric Synthesis*, Wiley-VCH, Weinheim, Germany, 2005.
21. Lelais, G. MacMillan, D. W. C. *Aldrichim. Acta* 2006, *39*, 78–87.
22. Gaunt, M. J.; Johansson, C. C. C.; McNally, A.; Vo, N. T. *Drug Discov. Today* 2007, *12*, 8–27.
23. *Chem. Rev.* 2007, *107*, 5413–5883.
24. MacMillan, D. W. C. *Nature* 2008, *455*, 304–308.
25. Pellissier, H. *Recent Developments in Asymmetric Organocatalysis*, Royal Society of Chemistry, Cambridge, UK, 2010.
26. List, B.; Maruoka, K., Eds. *Science of Synthesis: Asymmetric Organocatalysis*, Thieme, Stuttgart, Germany, 2012.
27. Dalko, I., Ed. *Comprehensive Enantioselective Organocatalysis: Catalysts, Reactions, and Applications*, Wiley-VCH, Weinheim, Germany, 2013.
28. Anastas, P. T.; Warner, J. C. *Green Chemistry: Theory and Practice*, Oxford University Press, New York, 1998.
29. Dunn, P. J.; Wells, A. S.; Williams, M. T., Eds. *Green Chemistry in the Pharmaceutical Industry*, Wiley-VCH, Weinheim, Germany, 2010.
30. Jiménez-González, C.; Constable, D. J. C. *Green Chemistry and Engineering: A Practical Design Approach*, Wiley, Hoboken, NJ, 2011.
31. Zhang, W.; Cue, B. W., Jr., Eds. *Green Techniques for Organic Synthesis and Medicinal Chemistry*, Wiley, Chichester, 2012.
32. Anastas, P. T.; Warner, J. C. *Green Chemistry: Theory and Practice*, Oxford University Press, New York, 1998, p. 30.
33. Huang, Y.; Walji, A. M.; Larsen, C. H.; MacMillan, D. W. C. *J. Am. Chem. Soc.* 2005, *127*, 15051–15053.
34. Beeson, T. D.; Mastracchio, A.; Hong, J.-B.; Ashton, K.; MacMillan, D. W. C. *Science* 2007, *316*, 582–585.
35. Walji, A. M.; MacMillan, D. W. C. *Synlett* 2007, 1477–1489.
36. Sibi, M. P.; Hasegawa, M. *J. Am. Chem. Soc.* 2007, *129*, 4124–4125.
37. Jang, H.-Y.; Hong, J.-B.; MacMillan, D. W. C. *J. Am. Chem. Soc.* 2007, *129*, 7004–7005.
38. Nicewicz, D. A.; MacMillan, D. W. C. *Science* 2008, *322*, 77–80.
39. Kim, H.; MacMillan, D. W. C. *J. Am. Chem. Soc.* 2008, *130*, 398–399.
40. Graham, T. H.; Jones, C. M.; Jui, N. T.; MacMillan, D. W. C. *J. Am. Chem. Soc.* 2008, *130*, 16494–14695.
41. Nicolaou, K. C.; Reingruber, R.; Sarlah, D.; Bräse, S. *J. Am. Chem. Soc.* 2009, *131*, 2086–2087.
42. Nicolaou, K. C.; Reingruber, R.; Sarlah, D.; Bräse, S. *J. Am. Chem. Soc.* 2009, *131*, 6640.
43. Simmons, B.; Walji, A. M.; MacMillan, D. W. C. *Angew. Chem. Int. Ed.* 2009, *48*, 4349–4353.
44. Amatore, M.; Beeson, T. D.; Brown, S. P.; MacMillan, D. W. C. *Angew. Chem. Int. Ed.* 2009, *48*, 5121–5124.
45. Nagib, D. A.; Scott, M. E.; MacMillan, D. W. C. *J. Am. Chem. Soc.* 2009, *131*, 10875–10877.
46. Wilson, J. E.; Casarez, A. D.; MacMillan, D. W. C. *J. Am. Chem. Soc.* 2009, *131*, 11332–11334.
47. Conrad, J. C.; Kong, J.; Laforteza, B. N. MacMillan, D. W. C. *J. Am. Chem. Soc.* 2009, *131*, 11640–11641.
48. Rendler, S.; MacMillan, D. W. C. *J. Am. Chem. Soc.* 2010, *132*, 5027–5029.

49. Jui, N. T.; Lee, E. C. Y.; MacMillan, D. W. C. *J. Am. Chem. Soc.* 2010, *132*, 10015–10017.
50. MacMillan, D. W. C.; Beeson, T. D. SOMO and radical chemistry in organocatalysis. In *Science of Synthesis: Asymmetric Organocatalysis 1*, List, B., Ed.; Thieme, Stuttgart, Germany, 2012, pp. 271–307.
51. Han, Z.-Y.; Wang, C.; Gong, L.-Z. Organocatalysis combined with metal catalysis or biocatalysis. In *Science of Synthesis: Asymmetric Organocatalysis 2*, Maruoka, K., Ed.; Thieme, Stuttgart, Germany, 2012, pp. 697–740.
52. Chen, Y.-C.; Cui, H.-L. Organocatalytic cascade reactions. In *Science of Synthesis: Asymmetric Organocatalysis 2*, Maruoka, K., Ed.; Thieme, Stuttgart, Germany, 2012, pp. 787–845.
53. Simonovich, S. P.; Van Humbeck, J. F.; MacMillan, D. W. C. *Chem. Sci.* 2012, *3*, 58–61.
54. Skucas, E.; MacMillan, D. W. C. *J. Am. Chem. Soc.* 2012, *134*, 9090–9093.
55. Jui, N. T.; Garber, J. A. O.; Finelli, F. G.; MacMillan, D. W. C. *J. Am. Chem. Soc.* 2012, *134*, 11400–11403.
56. Comito, R. J.; Finelli, F. G.; MacMillan, D. W. C. *J. Am. Chem. Soc.* 2013, *135*, 9358–9361.
57. Laforteza, B. N.; Pickworth, M.; MacMillan, D. W. C. *Angew. Chem. Int. Ed.* 2013, *52*, 11269–11272.
58. Cecere, G.; König, C. M.; Alleva, J. L.; MacMillan, D. W. C. *J. Am. Chem. Soc.* 2013, *135*, 11521–11524.
59. Stevens, J. M.; MacMillan, D. W. C. *J. Am. Chem. Soc.* 2013, *135*, 11756–11759.
60. List, B. *Acc. Chem. Res.* 2004, *37*, 548–557.
61. Mukherjee, S.; Yang, J. W.; Hoffmann, S.; List, B. *Chem. Rev.* 2007, *107*, 5471–5569.
62. Erkkilä, A.; Majander, I.; Pihko, P. M. *Chem. Rev.* 2007, *107*, 5416–5470.
63. MacMillan, D. W. C.; Watson, A. J. B. α-Functionalization of carbonyl compounds. In *Science of Synthesis: Stereoselective Synthesis 3*, Evans, P. A., Ed.; Thieme, Stuttgart, Germany, 2011, pp. 675–745.
64. Brazier, J. B.; Tomkinson, N. C. O. *Top. Curr. Chem.* 2010, *291*, 281–347.
65. Wang, X.-W.; Wang, Y.; Jia, J. Enamine catalysis of intramolecular aldol reactions. In *Science of Synthesis: Asymmetric Organocatalysis 1*, List, B., Ed.; Thieme, Stuttgart, Germany, 2012, pp. 1–33.
66. Wang, X.-W.; Wang, Y.; Jia, J. Enamine catalysis of intermolecular aldol reactions. In *Science of Synthesis: Asymmetric Organocatalysis 1*, List, B., Ed.; Thieme, Stuttgart, Germany, 2012, pp. 35–72.
67. Benohoud, M.; Hayashi, Y. Enamine catalysis of Mannich reactions. In *Science of Synthesis: Asymmetric Organocatalysis 1*, List, B., Ed.; Thieme, Stuttgart, Germany, 2012, pp. 73–134.
68. Mase, M. Enamine catalysis of Michael reactions. In *Science of Synthesis: Asymmetric Organocatalysis 1*, List, B., Ed.; Thieme, Stuttgart, Germany, 2012, pp. 135–216.
69. Mukherjee, S. Enamine catalysis of α-functionalizations and alkylations. In *Science of Synthesis: Asymmetric Organocatalysis 1*, List, B., Ed.; Thieme, Stuttgart, Germany, 2012, pp. 217–270.
70. MacMillan, D. W. C.; Watson, A. J. B. Iminium catalysis. In *Science of Synthesis: Asymmetric Organocatalysis 1*, List, B., Ed.; Thieme, Stuttgart, Germany, 2012, pp. 309–401.
71. Liu, Y.; Melchiorre, P. Iminium catalysis with primary amines. In *Science of Synthesis: Asymmetric Organocatalysis 1*, List, B., Ed.; Thieme, Stuttgart, Germany, 2012, pp. 403–438.
72. Northrup, A. B.; MacMillan, D. W. C. *J. Am. Chem. Soc.* 2002, *124*, 2458–2460.
73. Jensen, K. L.; Dickmeiss, G.; Jiang, H.; Albrecht, Ł.; Jørgensen, K. A. *Acc. Chem. Res.* 2012, *45*, 248–264.

74. Li, J.-L.; Liu, T.-Y.; Chen, Y.-C. *Acc. Chem. Res.* 2012, *45*, 1491–1500.
75. Arceo, E.; Melchiorre, P. *Angew. Chem. Int. Ed.* 2012, *51*, 5290–5292.
76. Ramachary, D. B.; Reddy, Y. V. *Eur. J. Org. Chem.* 2012, 865–887.
77. Kumar, I.; Ramaraju, P.; Mir, N. A. *Org. Biomol. Chem.* 2013, *11*, 709–716.
78. Jiang, H.; Albrecht, Ł.; Jørgensen, K. A. *Chem. Sci.* 2013, *4*, 2287–2300.
79. Jurberg, I. D.; Chatterjee, I.; Tannert, R.; Melchiorre, P. *Chem. Commun.* 2013, *49*, 4869–4883.
80. Wurz, R. P. *Chem. Rev.* 2007, *107*, 5570–5595.
81. Müller, C. E.; Schreiner, P. R. *Angew. Chem. Int. Ed.* 2011, *50*, 6012–6042.
82. Furuta, T.; Kawabata, T. Chiral DMAP-type catalysts for acyl transfer reactions. In *Science of Synthesis: Asymmetric Organocatalysis 1*, List, B., Ed.; Thieme, Stuttgart, Germany, 2012, pp. 497–546.
83. Smith, A. D.; Woods, P. A. Non-DMAP-type catalysts for acyl-transfer reactions. In *Science of Synthesis: Asymmetric Organocatalysis 1*, List, B., Ed.; Thieme, Stuttgart, Germany, 2012, pp. 547–590.
84. Enríquez-García, Á.; Kündig, E. P. *Chem. Soc. Rev.* 2012, *41*, 7803–7831.
85. Enders, D.; Niemeier, O.; Henseler, A. *Chem. Rev.* 2007, *107*, 5606–5655.
86. Suzuki, K.; Takikawa, H. Carbene-catalyzed benzoin reactions. In *Science of Synthesis: Asymmetric Organocatalysis 1*, List, B., Ed.; Thieme, Stuttgart, Germany, 2012, pp. 591–618.
87. DiRocco, D. A.; Rovis, T. Carbene-catalyzed Stetter reactions In *Science of Synthesis: Asymmetric Organocatalysis 1*, List, B., Ed.; Thieme, Stuttgart, Germany, 2012, pp. 619–637.
88. Chiang, P.-C.; Bode, J. W. N-Heterocyclic carbene catalyzed reactions of α-functionalized aldehydes. In *Science of Synthesis: Asymmetric Organocatalysis 1*, List, B., Ed.; Thieme, Stuttgart, Germany, 2012, pp. 639–672.
89. Bugaut, X.; Glorius, F. *Chem. Soc. Rev.* 2012, *41*, 3511–3522.
90. Hatakeyama, S. (Aza)-Morita-Baylis-Hillman reactions. In *Science of Synthesis: Asymmetric Organocatalysis 1*, List, B., Ed.; Thieme, Stuttgart, Germany, 2012, pp. 673–721.
91. Wei, Y.; Shi, M. *Chem. Rev.* 2013, *113*, 6659–6690.
92. Chen, S.; Salo, E. C.; Kerrigan, N. J. Tertiary amine and phosphine-catalyzed reactions of ketenes and α-halo ketones. In *Science of Synthesis: Asymmetric Organocatalysis 1*, List, B., Ed.; Thieme, Stuttgart, Germany, 2012, pp. 455–496.
93. Fan, Y. C.; Kwon, O. Phosphine catalysis. In *Science of Synthesis: Asymmetric Organocatalysis 1*, List, B., Ed.; Thieme, Stuttgart, Germany, 2012, pp. 723–782.
94. Wang, Z.; Xu, X.; Kwon, O. *Chem. Soc. Rev.* 2014, *43*, 2927–2940.
95. Yang, D. *Acc. Chem. Res.* 2004, *37*, 497–505.
96. Wong, O. A.; Ramirez, T. A.; Shi, Y. Asymmetric ketone and iminium salt catalyzed epoxidations. In *Science of Synthesis: Asymmetric Organocatalysis 1*, List, B., Ed.; Thieme, Stuttgart, Germany, 2012, pp. 783–830.
97. Akiyama, T. *Chem. Rev.* 2007, *107*, 5744–5758.
98. Zamfir, A.; Schenker, S.; Freund, M.; Tsogoeva, S. B. *Org. Biomol. Chem.* 2010, *8*, 5262–5276.
99. Akiyama, T. Phosphoric acid catalyzed reactions of imines organocatalysts. In *Science of Synthesis: Asymmetric Organocatalysis 2*, Maruoka, K., Ed.; Thieme, Stuttgart, Germany, 2012, pp. 169–217.
100. Terada, M.; Momiyama, N. Phosphoric acid catalysis of reactions not involving imines organocatalysts. In *Science of Synthesis: Asymmetric Organocatalysis 2*, Maruoka, K., Ed.; Thieme, Stuttgart, Germany, 2012, pp. 219–278.
101. Hashimoto, T. Brønsted acid catalysts other than phosphoric acids. In *Science of Synthesis: Asymmetric Organocatalysis 2*, Maruoka, K., Ed.; Thieme, Stuttgart, Germany, 2012, pp. 279–296.
102. Mahlau, M.; List, B. *Angew. Chem. Int. Ed.* 2013, *52*, 518–533.

103. Brak, K.; Jacobsen, E. N. *Angew. Chem. Int. Ed.* 2013, *52*, 534–561.
104. Jiang, L.; Chen, Y.-C. *Catal. Sci. Technol.* 2011, *1*, 354–365.
105. Singh, R. P.; Deng, L. Cinchona alkaloid organocatalysts. In *Science of Synthesis: Asymmetric Organocatalysis 2*, Maruoka, K., Ed.; Thieme, Stuttgart, Germany, 2012, pp. 41–117.
106. Gröger, H. *Chem. Rev.* 2003, *103*, 2795–2827.
107. Leow, D.; Tan, C.-H. *Chem. Asian. J.* 2009, *4*, 488–507.
108. Nagasawa, K.; Sohtome, Y. Chiral guanidine and amidine organocatalysts. In *Science of Synthesis: Asymmetric Organocatalysis 2*, Maruoka, K., Ed.; Thieme, Stuttgart, Germany, 2012, pp. 1–40.
109. Taylor, J. E.; Bull, S. D.; Williams, J. M. J. *Chem. Soc. Rev.* 2012, *41*, 2109–2121.
110. Taylor, M. S.; Jacobsen, E. N. *Angew. Chem. Int. Ed.* 2006, *45*, 1520–1543.
111. Doyle, A. G.; Jacobsen, E. N. *Chem. Rev.* 2007, *107*, 5713–5743.
112. Hof, K.; Lippert, K. M.; Schreiner, P. R. Hydrogen-bonding catalysts: (Thio)urea catalysis. In *Science of Synthesis: Asymmetric Organocatalysis 2*, Maruoka, K., Ed.; Thieme, Stuttgart, Germany, 2012, pp. 297–412.
113. Uraguchi, D.; Ooi, T. Hydrogen-bonding catalysts other than ureas and thioureas. In *Science of Synthesis: Asymmetric Organocatalysis 2*, Maruoka, K., Ed.; Thieme, Stuttgart, Germany, 2012, pp. 413–435.
114. Paull, D. H.; Abraham, C. J.; Scerba, M. T.; Alden-Danforth, E.; Leckta, T. *Acc. Chem. Res.* 2008, *41*, 655–663.
115. Siau, W.-Y.; Wang, J. *Catal. Sci. Technol.* 2011, *1*, 1298–1310.
116. Jang, H. B.; Oh, J. S.; Song, C. E. Bifunctional cinchona alkaloid organocatalysts. In *Science of Synthesis: Asymmetric Organocatalysis 2*, Maruoka, K., Ed.; Thieme, Stuttgart, Germany, 2012, pp. 119–168.
117. Inokuma, T.; Takemoto, Y. Bifunctional (Thio)urea and BINOL catalysts. In *Science of Synthesis: Asymmetric Organocatalysis 2*, Maruoka, K., Ed.; Thieme, Stuttgart, Germany, 2012, pp. 437–497.
118. Ooi, T.; Maruoka, K. *Angew. Chem. Int. Ed.* 2007, *46*, 4222–4266.
119. Hashimoto, T.; Maruoka, K. *Chem. Rev.* 2007, *107*, 5656–5682.
120. Park, H.-G. Phase-transfer catalysis: Natural-product-derived PTC. In *Science of Synthesis: Asymmetric Organocatalysis 2*, Maruoka, K., Ed.; Thieme, Stuttgart, Germany, 2012, pp. 499–549.
121. Shirakawa, S.; Maruoka, K. Phase transfer catalysis: Non-natural-product-derived PTC. In *Science of Synthesis: Asymmetric Organocatalysis 2*, Maruoka, K., Ed.; Thieme, Stuttgart, Germany, 2012, pp. 551–599.
122. Shirakawa, S.; Maruoka, K. *Angew. Chem. Int. Ed.* 2013, *52*, 4312–4348.
123. Trost, B. M. *Science* 1991, *254*, 1471–1477.
124. Sheldon, R. A. *CHEMTECH* 1994, 38–47.
125. Hudlicky, T.; Koroniak, D. A.; Claeboe, C. D.; Brammer, L. E. *Green Chem.* 1999, *1*, 57–59.
126. Curzons, A. D.; Constable, D. J. C.; Mortimer, D. N.; Cunningham, V. L. *Green Chem.* 2001, *3*, 1–6.
127. Guineé, J. B., Ed. *Handbook on Life Cycle Assessment: Operational Guide to the ISO Standards*, Kluwer Academic Publishers, Dordrecht, 2002.
128. Constable, D. J. C.; Curzons, A. D.; Cunningham, V. L. *Green Chem.* 2002, *4*, 521–527.
129. Gronnow, M. J.; White, R. J.; Clark, J. H.; Macquarrie, D. J. *Org. Process. Res. Dev.* 2005, *9*, 516–518.
130. Andraos, J. *Org. Process Res. Dev.* 2005, *9*, 149–163.
131. Andraos, J. *Org. Process Res. Dev.* 2005, *9*, 404–431.
132. Sheldon, R. A. *Green Chem.* 2007, *9*, 1273–1283.
133. Lapkin, A.; Constable, D. J. C. *Green Chemistry Metrics: Measuring and Monitoring Sustainable Processes*, Wiley, Chichester, 2008.

134. Clavo-Flores, F. G. *ChemSusChem* 2009, *2*, 905–919.
135. Mulvihill, M. J.; Beach, E. S.; Zimmerman, J. B.; Anastas, P. T. *Annu. Rev. Environ. Resour.* 2011, *36*, 271–293.
136. Jiménez-González, C.; Ponder, C. S.; Broxterman, Q. B.; Manley, J. B. *Org. Process Res. Dev.* 2011, *15*, 912–917.
137. Watson, W. J. W. *Green Chem.* 2012, *14*, 251–259.
138. Augé, J.; Scherrman, M.-C. *New J. Chem.* 2012, *36*, 1091–1098.
139. Sheldon, R. A. *Chem. Soc. Rev.* 2012, *41*, 1437–1451.
140. Dunn, P. J. *Chem. Soc. Rev.* 2012, *41*, 1452–1461.
141. Jiménez-González, C.; Constable, D. J. C.; Ponder, C. S. *Chem. Soc. Rev.* 2012, *41*, 1485–1498.
142. Ruiz-Mercado, G. J.; Smith, R. L.; Gonzalez, M. A. *Ind. Eng. Chem. Res.* 2012, *51*, 2329–2353.
143. Ruiz-Mercado, G. J.; Smith, R. L.; Gonzalez, M. A. *Ind. Eng. Chem. Res.* 2012, *51*, 2309–2328.
144. Ribeiro, M. G. T. C.; Machado, A. A. S. C. *Green Chem. Lett. Rev.* 2013, *6*, 1–18.
145. McGonagle, F. I.; Sneddon, H. F.; Jamieson, C.; Watson, A. J. B. *ACS Sustainable Chem. Eng.* 2014, *2*, 523–532.
146. Constable, D. J. C.; Jiménez-González, C.; Henderson, R. K. *Org. Process. Res. Dev.*, 2007, *11*, 133–137.
147. Alfonsi, K.; Colberg, J.; Dunn, P. J.; Fevig, T.; Jennings, S.; Johnson, T. A.; Kleine, H. P. et al. *Green Chem.* 2008, *10*, 31–36.
148. Cue, B. W.; Zhang, J. *Green Chem. Lett. Rev.*, 2009, *2*, 193–211.
149. Henderson, R. K.; Jiménez-González, C.; Constable, D. J. C.; Alston, S. R.; Inglis, G. G. A.; Fisher, G.; Sherwood, J.; Binks, S. P.; Curzons, A. D. *Green Chem.* 2011, *13*, 854–862.
150. MacMillan, D. S.; Murray, J.; Sneddon, H. F.; Jamieson, C.; Watson, A. J. B. *Green Chem.* 2012, *14*, 3016–3019.
151. Taygerly, J. P.; Miller, L. M.; Yee, A.; Peterson, E. A. *Green Chem.* 2012, *14*, 3020–3025.
152. MacMillan, D. S.; Murray, J.; Sneddon, H. F.; Jamieson, C.; Watson, A. J. B. *Green Chem.* 2013, *15*, 596–600.
153. McGonagle, F. I.; MacMillan, D. S.; Murray, J.; Sneddon, H. F.; Jamieson, C.; Watson, A. J. B. *Green Chem.* 2013, *15*, 1159–1165.
154. Raj, M.; Singh, V. K. *Chem. Commun.* 2009, 6687–6703.
155. Neher, R. In *Thin Layer Chromatography*, Marini-Bettolo, G. B., Ed.; Elsevier, Amsterdam, 1964, p. 77.
156. Yoshikawa, N.; Yamada, Y. M. A.; Das, J.; Sasai, H.; Shibasaki, M. *J. Am. Chem. Soc.* 1999, *121*, 4168–4178.
157. Zuend, S. J.; Coughlin, M. P.; Lalonde, M. P.; Jacobsen, E. N. *Nature* 2009, *461*, 968–970.
158. Hatano, M.; Hattori, Y.; Furuya, Y.; Ishihara, K. *Org. Lett.* 2009, *11*, 2321–2324.
159. Brandau, S.; Landa, A.; Franzén, J.; Marigo, M.; Jørgensen, K. A. *Angew. Chem. Int. Ed.* 2006, *45*, 4305–4309.
160. Velmthi, S.; Swarnalakshmi, S.; Narasimhan, S. *Tetrahedron: Asymm.* 2003, *14*, 113–117.
161. Nakashima, D.; Yamamoto, H. *J. Am. Chem. Soc.* 2006, *128*, 9626–9627.
162. Liu, D.; Canales, E.; Corey, E. J. *J. Am. Chem. Soc.* 2007, *129*, 1498–1499.
163. Wang, X.; List, B. *Angew. Chem. Int. Ed.* 2008, *47*, 1119–1122.
164. Payne, G. B. *J. Org. Chem.* 1960, *25*, 275–276.
165. Brown, S. P.; Brochu, M. P.; Sinz, C. J.; MacMillan, D. W. C. *J. Am. Chem. Soc.* 2003, *125*, 10808–10809.

166. Beshara, C. S.; Hall, A.; Jenkins, R. L.; Jones, K. L.; Jones, T. C.; Killeen, N. M.; Taylor, P. H.; Thomas, S. P.; Tomkinson, N. C. O. *Org. Lett.* 2005, *7*, 5729–5732.
167. Information on the current supply and projected future supply of elements can be obtained from the British Geological Survey. Available at http://www.bgs.ac.uk/mineralsuk/home.html.
168. Bryan, M. C.; Dillon, B.; Hamann, L. G.; Hughes, G. J.; Kopach, M. E.; Peterson, E. A.; Pourashraf, M. et al. *J. Med. Chem.* 2013, *56*, 6007–6021.
169. Hernández, J. G.; Juaristi, E. *Chem. Commun.* 2012, *48*, 5396–5409.

4 Learning from Biology
Biomimetic Catalysis

Thomas P. Umile and Ryan S. Buzdygon

CONTENTS

4.1 ENZYMES

Nature has evolved a host of proteins that catalyze the chemical reactions necessary to sustain life. These biological catalysts are called enzymes, and a given enzyme catalyzes a specific reaction (or subset of reactions). In addition to enhancing the rates of the reactions by orders of magnitudes, enzymes must conform to a strict set of operating conditions. Enzymes must operate in the cell's complex matrix of biomolecules, select for only the proper substrate on which to act, function under the ambient conditions of the organism in which it exists, and be constructed of and use the simplest and most abundant materials possible. Due to enzymes' tremendous activity despite these narrow operational criteria, there is significant interest in understanding the chemistry of enzymes. Furthermore, tapping the benefits of these biological catalysts (biocatalysts) for use in the production of commodity chemicals is accordingly an area of active research, and these efforts include not only the direct use of enzymes but also the development of synthetic catalysts that mimic enzymes' most desirable activities.

Enzymes, like all proteins, are composed of amino acids. These amino acids link together in long polymeric chains that comprise the enzyme's primary structure (Figure 4.1). Median chain lengths for eukaryotic and bacterial proteins are 361 and 267 amino acids, respectively.[1] These amino acid chains coil, fold, and aggregate into precise three-dimensional shapes. Together, the defined shape and amino acid composition of an enzyme defines its properties and activity. Some enzymes additionally incorporate cofactors or prosthetic groups, which are non–amino acid molecules that provide additional functionality. Broadly speaking, an enzyme catalyzes a reaction by binding to a target substrate molecule, orienting the compound in a specific way, and placing it in the proximity of some highly reactive functional group(s) on the enzyme where the actual reaction occurs. The location on (or in) the enzyme where the catalysis occurs is commonly called its active site.

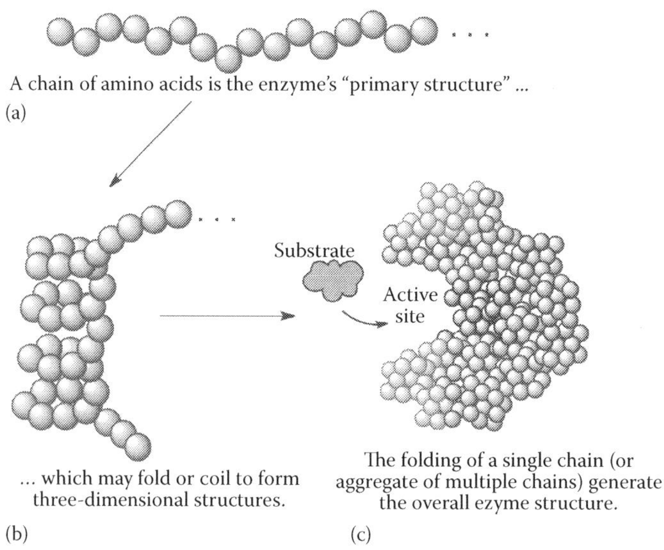

FIGURE 4.1 Amino acids (represented as spheres) link together in chains (a) that fold or coil (b) to form larger three-dimensional structures (c). A substrate molecule binds at the active site where the catalyzed reaction occurs.

TABLE 4.1
Reaction Types Catalyzed by Enzymes

Enzyme Class	Reaction Type Catalyzed
Oxidoreductase	Oxidation and reduction (electron transfer)
Transferase	Transfer of a functional group from one molecule to another
Hydrolase	Cleavage of covalent bonds by the addition of water
Lyase	Elimination of a small molecule for a double bond or cyclic structure
Isomerase	Rearrangement of the atoms in a substrate
Ligase	Covalently link two smaller molecules

4.1.1 Nature's Remarkable Catalysts

The International Union of Biochemistry and Molecular Biology classifies enzymes into six families that describe the fundamental reaction types they each catalyze (Table 4.1).[2] Many of the chemical transformations required for commodity or fine chemical production are represented within these classes, including oxidation and reduction, the functionalization of C–H bonds, and C–C bond formation.[3,4]

A few notable characteristics are typical of enzyme-catalyzed reactions. These features make enzymes crucial to biological function and also highly desirable as catalysts for synthetic or industrial applications. They include dramatic rate enhancements, exquisite substrate-level selectivity, reaction specificity, efficient operation under mild reaction conditions, and the use of simple and abundant materials and "reagents."

4.1.1.1 Dramatic Rate Enhancements

Enzymes, as catalysts, enhance the rate of chemical reactions. The range of efficiencies is broad, and rate enhancements (compared with the uncatalyzed reaction) can range from 10^7-fold to 10^{19}-fold![5] For example, the conversion of dihydroxyacetone phosphate (DHAP) to glyceraldehyde-3-phosphate (GAP; Figure 4.2a) is catalyzed by triosephosphate isomerase with a rate enhancement of 1.0×10^9 (essentially decreasing the half-life of the uncatalyzed reaction from 1.9 days to less than 0.1 ms).[6–8] For some enzymes, like triosephosphate isomerase, the rate of the catalyzed reaction is actually limited only by the diffusion of the reactant and enzyme molecules.[9] In effect, only the speed with which the substrate and enzyme encounter one another limits the catalytic rate in this case.

A full description of the chemical basis for enzymatic rate enhancement is beyond the scope of this chapter, and the reader is referred to thoughtful reviews and discussions on the matter.[5,11–13] However, all explanations for enzymatic catalysis recognize the importance of the active site architecture. In the active site, a bound substrate is brought in the proximity of the reactive amino acid functionalities (or prosthetic groups), increasing the "effective concentration" of these species and their propensity to react with each other.[12] Simultaneously, other amino acids or cofactors can direct the conformation of a substrate or stabilize some transition state structure through

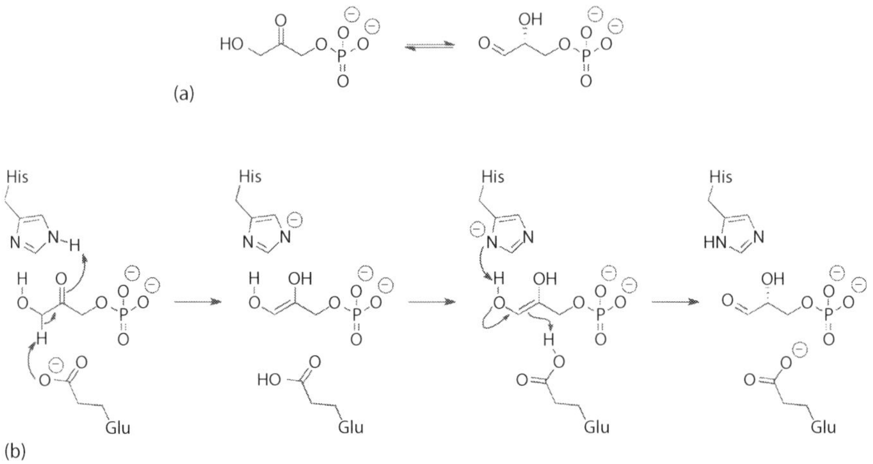

FIGURE 4.2 (a) Isomerization of DHAP to GAP; (b) mechanism of DHAP isomerization in the active site of triosephosphate isomerase.[10]

noncovalent interactions. The exact positioning of amino acid residues or other functionalities within the active site is a major factor that provides for a remarkable reaction rate enhancement. In the case of triosephosphate isomerase, basic glutamate and acidic histidine* residues in the active site are precisely located to shuttle protons and facilitate the isomerization reaction of DHAP (Figure 4.2b).[10]

4.1.1.2 Substrate Selectivity

In the laboratory, the chemist or engineer controls a reaction through careful inclusion (or exclusion) of specific reagents added to a flask in timed and precise ways. The enzyme, by contrast, exercises its selectivity only by permitting a particular substrate (or family of substrates) to access its active site; those molecules that can approach the reactive functionality buried within will react. (In those cases where more than one substrate or reagent must be bound by the enzyme, the order in which the compounds are bound is sometimes carefully orchestrated by the enzyme.) The size of a substrate obviously plays a role in determining if a molecule can access the active site, but more often the architecture of the active site is such that other noncovalent factors such as the shape or polarity of the site are ideal to bind only the "correct" substrate.[14] Enzyme–substrate recognition can sometimes be so exact that an early conceptualization compared it to a "lock and key."[15]

4.1.1.3 Reaction Specificity and Selectivity

Related to substrate selectivity, the chemical reactions catalyzed by enzymes often display a reaction specificity or selectivity: to react with one part of a given substrate over another, produce a specific stereoisomer over another, or react with a particular functional group over another. This control arises from such factors as the conformation-specific binding of a substrate to the active site and the precise mechanism by which

* Interestingly, the imidazole of histidine is *not* protonated in the active site of the enzyme.

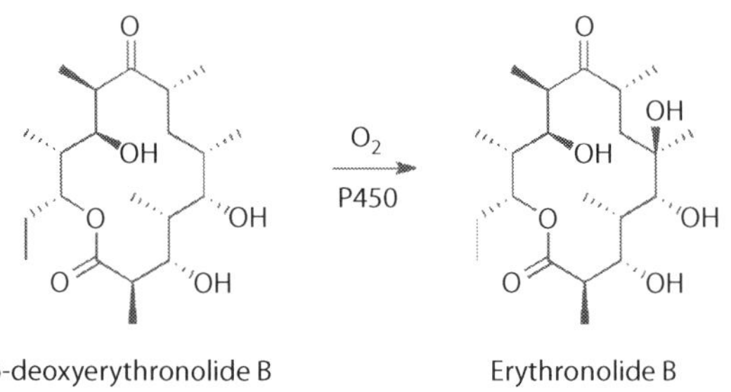

FIGURE 4.3 Cytochrome P450 enzyme, eryF, catalyzes the aerobic oxidation of 6-deoxyerythronolide, with retention of stereochemistry. (From Staunton, J., and Wilkinson, B. *Chem Rev* 1997, 97 (7), 2611–2629.)

the catalyzed reaction occurs. It is particularly intriguing when the enzyme displays exquisite control to react with a specific functional group or portion of the substrate even in the presence of more reactive structural features. For example, the enzyme eryF catalyzes the stereospecific hydroxylation of 6-deoxyerythronolide B exclusively at C-6, ignoring other tertiary C–H bonds or the more reactive alcohol functionalities (Figure 4.3).[16]

4.1.1.4 Ambient Conditions

As components of biological systems, enzymes must operate efficiently under the ambient conditions of the organism in which it resides, which may be narrow and normally less intensive than those commonly used in industrial systems. For example, most thermoregulated mammals have core body temperatures of 36°C to 38°C, and intracellular pH in eukaryotes is limited to 7.0 to 7.4 (precluding high temperatures and strongly acidic or basic reaction conditions).[17,18] Simultaneously, however, other organisms are adapted to survive under extreme conditions[19,20] [such as cryophiles,[21] which live in Antarctic seawaters (~1°C)], and remarkably the enzymes of these organisms are evolved to operate with similar efficiency under those very different conditions.

4.1.1.5 Use of Abundant Materials

Molecules built by nature are derived ultimately from the precursors that are abundant in the environment such as water, carbon dioxide (CO_2), oxygen (O_2), ammonia (NH_3), sulfur, and phosphates.[22] Most notably, plants use solar energy to fix the carbon in CO_2 as sugars that are eventually transformed into all of the fats, proteins, and other complex biomolecules of life (Figure 4.4). Although these processes involve a multitude of individual steps, most, if not at all, are catalyzed by enzymes. Nature sets an important precedent and example by using catalysts to convert simple and abundant materials into complex products.

Moreover, because proteins themselves are products of biology, enzymes are similarly constructed from relatively simple and abundant starting materials. Proteins are chiefly built from 20 common amino acids,[23] each of which provides a functional group to modulate protein intermolecular and intramolecular interactions and reactivity. Despite the relatively limited selection of components, however, the catalytic

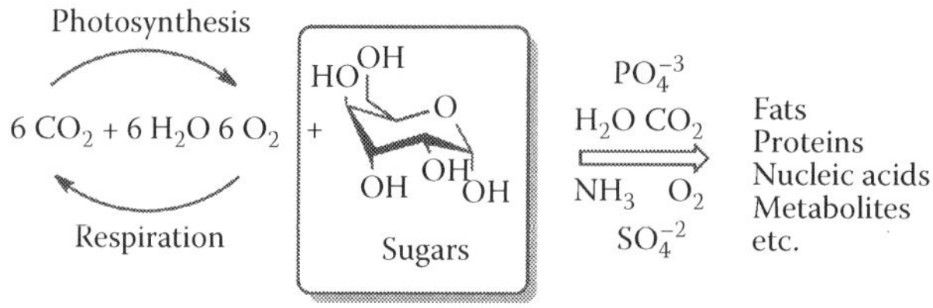

FIGURE 4.4 Photosynthesis fixes carbon from CO_2 in sugars, which are converted into all the biomolecules of life by complex metabolic pathways by enzymes using simple "reagents" that come ultimately from the geosphere.

efficiency of enzymes often rivals and bests that of synthetic catalysts. Moreover, any non–amino acid cofactors or prosthetic groups that the enzyme employs must be similarly constructed of materials readily found in the environment. Consider, for example, the metalloenzymes that require metal cofactors in their active sites. Unlike the synthetic chemist with access to expensive, precious metals such as platinum, palladium, and rhodium, metalloenzymes successfully employ metals that are abundant in the geosphere such as iron, manganese, and copper.[24] When an enzyme or other protein is damaged (resulting in a loss of activity or the development of some deleterious activity), the cell commonly disassembles the entire protein back to its constituent parts and rebuilds it anew.[25] This type of reuse and recycling capitalizes on nature's use of simple and abundant materials to generate structural complexity and thus desired activity.

4.2 BIOMIMETIC CATALYSTS

4.2.1 Inspired by Enzymes

Due to their activity and diverse capabilities, enzymes are already directly used in a variety of applications such as the production of commodity chemicals, bioremediation of pollution, and the synthesis of pharmaceutical compounds.[26] Where an enzyme is unknown or inefficient for a desired reaction, protein engineering techniques can be used to develop modified enzymes to meet the need.[27] (See Chapter 5 for further discussion on biocatalysis in the pharmaceutical industry.) Despite their fundamental benefits, however, industrial use of biocatalysts can be limited by such factors as their availability, instability toward necessary reaction conditions,[28] or the costs of producing them at the necessary scale.

A complementary approach for exploiting the characteristics of enzymes endeavors to develop *synthetic* catalysts that mimic those desired properties of enzymes, combining the robust nature of traditional chemical techniques with the enzymes' selectivity, speed, and other qualities. This tactic falls under the umbrella of *biomimicry*, which is a field focused on imitating the efficient and elegant designs of biology. It has been said that the goal of biomimicry is to learn from nature "what works" based on its 3.8 billion years of trial-and-error evolution.[29] As applied to chemistry and catalysis, biomimicry was first explored by Ronald Breslow at Columbia University for the directed stereoselective and regioselective oxidation of molecules such as steroids and long-chain alcohols.[30–32]

In the time since, biomimetic chemistry has grown into a multidisciplinary field that spans traditional organic, inorganic, and biochemistry and often requires expertise at the interface of these subfields. The practical goal to develop new chemical reactions and processes that mimic some aspect of biological systems includes not only enzymatic reactions and catalysis but also such pursuits as the structure and function of biological materials (e.g., spider silk) and the design of agricultural systems that mimic natural, self-sustaining ecosystems.[29]

4.2.2 Biomimetic Catalysts and Sustainability

Because enzymes can catalyze reactions that remain near impossibilities in the laboratory, the development of biomimetic catalysts may make such transformations accessible to industry and (ultimately) society. However, even in the cases where viable alternatives are already known, biomimetic catalysts remain an alluring target because many of the qualities of enzymes align well with principles of green chemistry and sustainability. As discussed above, enzymes not only effectively enhance the rates of chemical reactions but also do so under ambient conditions and with simple, abundant materials.

To better recognize how enzymes provide inspiration for sustainable chemistry, let us briefly consider a specific example: the family of cytochrome P450 enzymes. The cytochrome P450s are a family of iron-containing enzymes that catalyze the transfer of an oxygen atom from O_2 to an organic substrate (S), producing an oxidized product (SO) and water (H_2O, as a by-product) (Equation 4.1).[33,34] During this reaction, two electrons are necessary and provided by the reduced biological cofactors nicotinamide adenine dinucleotide (NADH) or nicotinamide adenine dinucleotide phosphate (NADPH).

$$O_2 + 2\,e^- + 2\,H^+ + S \rightarrow SO + H_2O \qquad (4.1)$$

This mono-oxygenation can produce alcohols from alkanes and epoxides from alkenes. Thousands of P450 enzymes are known and have been found in a range of organisms including bacteria, fungi, plants, insects, and mammals, and they are crucial to the metabolism of pharmaceutical compounds by the liver.[35] The various P450 isoforms have different substrate specificities.[36] Although the various members of the P450 family share some structural similarities, by far the most important conserved feature in P450 is the cysteine-ligated heme (iron protoporphyrin IX) found in the active site of the enzyme (Figure 4.11a).[33,37] This heme cofactor, ubiquitous in biochemistry, produces a characteristic UV–Vis absorbance at 450 nm upon binding a molecule of carbon monoxide to its reduced form, which gives the enzyme its name ("pigment 450").[38,39]

The oxygen-atom transfer reaction catalyzed by P450 is of much interest. Practically speaking, a fast, efficient, and selective C–H functionalization reaction (i.e., the P450 reaction itself) is the "holy grail" of chemical catalysis,[40] and could be directly used for more effectively preparing complex chemicals (e.g., pharmaceuticals) through late-state, selective functionalization[41] or generating alternative fuels or chemical precursors (e.g., turning currently unused, waste methane into value-added methanol).[42] Additionally, P450s can demonstrate exquisite selectivity for oxygenating one particular C–H bond of a substrate over another (see Figure 4.3). Moreover, alcohols (i.e., the oxygenated products of P450 catalysis) are generally considered *more reactive* than the starting organic substrate itself (due to the weaker C–H bonds in the product),[43,44] yet the reactive product is not commonly subjected to further oxidation by the enzyme. In meeting these challenges, the P450s present an inspiring lesson for the development of a more sustainable oxidation catalyst.

Molecular oxygen is valued as a particularly "green" oxidant because it is abundant in the atmosphere, renewable, environmentally safe, and carries few practical risks to its use (especially if air alone is its source).[45] Practically, O_2 is a challenging oxidant to use because, although potent, it is difficult to activate and control. From a synthetic perspective, there is a potential downside of aerobic, P450-type mono-oxygenation reactions because of the need for two electrons (Equation 4.1), and thus the requirement for an additional reagent if used industrially. However, P450 can use the similarly green hydrogen peroxide (H_2O_2) as an oxidant to carry out the same mono-oxygenation reaction (*vide infra*) without the need for an input of electrons or reducing equivalents (Equation 4.2).

$$H_2O_2 + S \rightarrow SO + H_2O \tag{4.2}$$

Additional facets of P450 activity provide biomimetic inspiration. The by-product of the P450 oxidation reaction is water, a much more benign compound than the unwanted chlorine-containing salts or other products generated during industrial oxidations for bleaching and wastewater treatment (with such reagents as Cl_2 or ClO_2).[46,47] Iron, the keystone of P450 activity, is the fourth most abundant element in the Earth's crust and is recognized as a less toxic (and cheaper!) alternative to more common precious metal catalysts.[48] It would be a dramatic oversimplification to say that P450s are doing chemistry we need using only air and dirt, yet they are a lot closer to that ideal than we currently are with our harsh reagents and expensive precious metal catalysts.

If sustainability can be defined as "doing more with less,"[49] there may not be a better example of it than what we find in biology. The cytochrome P450 are not unique, and nature commonly uses simple, clean, and abundant materials to carry out many reactions that society already uses or desperately needs. Importantly, "simple, clean, and abundant" can often translate to fewer environmental stresses and economic costs as well. The successful development of biomimetic catalysts can then simultaneously solve chemical challenges while addressing the needs of the triple bottom line of sustainability.

4.3 DEVELOPING BIOMIMETIC CATALYSTS

Mimicking the desirable features of enzymes is the fundamental goal in developing biomimetic catalysis. Key to the development of a successful biomimetic catalyst is a precise and detailed understanding of the chemical and physical properties that give rise to the enzymatic function to be imitated. Accordingly, the field is quite interdisciplinary and operates at the interface of organic, inorganic, and biochemistry. In developing a biomimetic catalyst, one may focus on capturing some principle of enzymatic function (such as the selective binding of substrate) or directly modeling the structure of the desired enzyme's active site (such a particularly intriguing reactive intermediate).

4.3.1 Mimicking Substrate Binding

4.3.1.1 Cyclodextrins

The selective binding of a substrate to an active site has been successfully mimicked in a number of ways. One of the earliest methods uses particularly simple but effective cyclodextrins,[50] which bind to some molecules in a process reminiscent of the original "lock-and-key" hypothesis of enzyme–substrate binding.[51] Cyclodextrins (Figure 4.5) are cyclic oligomers of glucose with a highly hydrophilic exterior.[52] The hydrophilic exterior facilitates the solubility of cyclodextrins in polar solvents (especially water), whereas the less-hydrophilic inner cavity provides an attractive binding site for less polar molecules.[52,53]

Without reactive functionalities, cyclodextrins normally act as little more than toroidal containers for encapsulating a molecule. However, even that feature alone can have a powerful effect on a reaction. For example, the encounter of diene and dienophile components of a Diels–Alder reaction is facilitated in the hydrophobic cavity of a cyclodextrin. The regioselectivity of the reaction between 1,3-pentadiene and 2,6-dimethylbenzoquinone is dramatically increased when the reaction is run in the presence of a stoichiometric amount of cyclodextrin (Figure 4.6).[54] Presumably, the two components encounter each other in the tight cavity of the cyclodextrin with a well-defined geometry.[55]

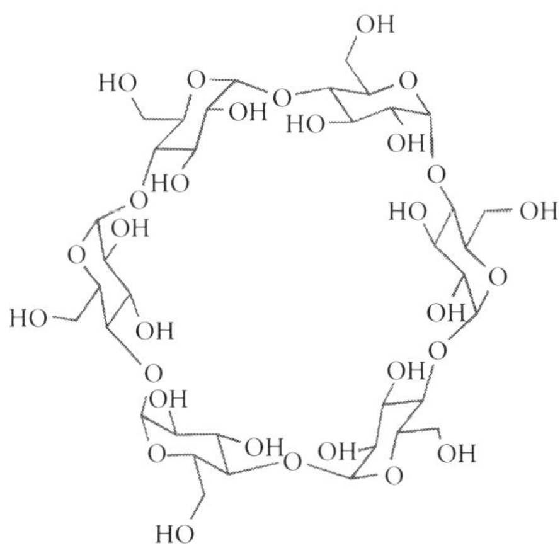

FIGURE 4.5 Structure of α-cyclodextrin. Common cyclodextrins consist of six to eight glucose monomers.

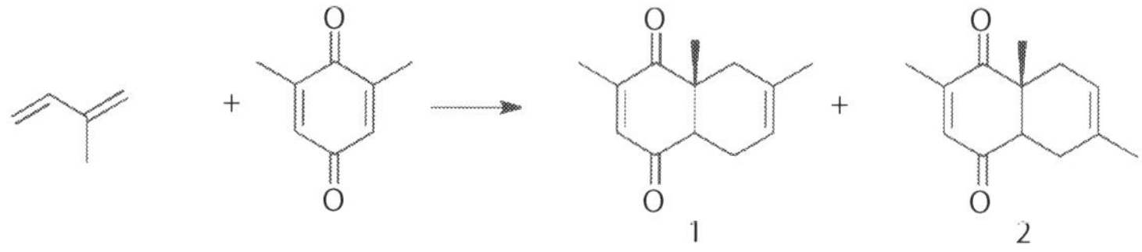

Conditions	Ratio 1:2	% yield
Water (8 d, room temperature)	65:35	4
Water/cyclodextrin (48 h, room temperature)	12:88	76

FIGURE 4.6 The regioselectivity Diels–Alder reaction is enhanced in the presence of a cyclodextrin. (Data from Chung, W.S., and Wang, J.Y. *J Chem Soc Chem Commun* 1995 (9), 971–972.)

4.3.1.2 Nanoreactors

In addition to simple cyclodextrins, other "hosts" capable of binding a substrate to enhance a reaction have been explored. The Raymond group has extensively studied the use of self-assembled "nanocages" or "nanoreactors" to enhance chemical reactions. These water-soluble "nanoreactors" (Figure 4.7) are composed of four metal(III) ions and six bis-bidentate catecholamide ligands (M_4L_6), which arrange into tetrahedral complexes and provide an interior cavity (300–500 Å) where "guest" molecules can bind.[56,57] Two strategies have been approached to apply these constructs to catalysis: encapsulate some reactive organometallic species into the cavity to facilitate its interaction with substrate *or* use the structure of the cavity itself as a catalyst by modulating some property of an encapsulated substrate.[56]

A thus-encapsulated iridium complex capable of C–H activation was demonstrated to discriminate among aldehyde substrates, as larger aldehydes presumably could not access the crowded cavity (akin to the substrate selectivity seen in enzymes).[57,58] More strikingly, however, the M_4L_6 complex itself can catalyze the rearrangement of a bound substrate by controlling its conformation or protonation state. The Ga_4L_6 host, for example, catalyzes the 3-aza-Cope rearrangement of

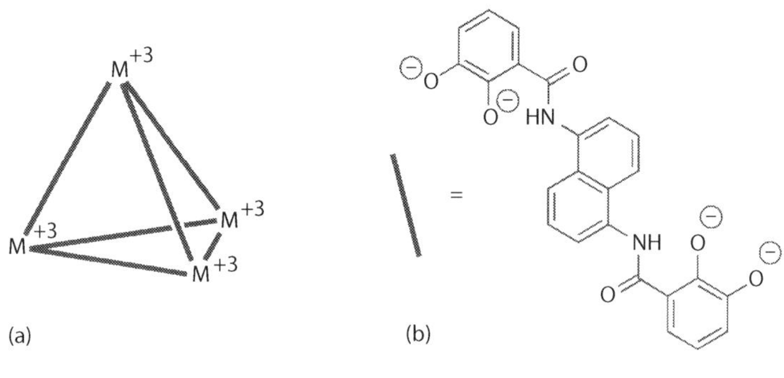

FIGURE 4.7 Tetrahedral "nanoreactors" (a) self-assemble from four metal(III) ions and six bis-bidentate catecholamide ligands (b).

encapsulated enammonium cations to γ,δ-unsaturated aldehydes (Figure 4.8).[59,60] Crowding in the cavity facilitates the rearrangement, presumably because the cavity enforces a conformation reminiscent of the chair-like transition state of the reaction. Modest rate accelerations compared with the uncatalyzed reaction (5–150×) were reported for most enammonum cations.[59] However, a particularly bulky substrate saw a rate enhancement of more than 800× because steric crowding in the host cavity forces the bulky alkenyl substituents closer than they would normally ever be in free solution.[59]

A particularly exciting result of the preference for the M_4L_6 complexes to bind cations is that the protonation of encapsulated *neutral* species is promoted. The effective pK_a for an encapsulated ammonium ion is shifted upward by as much as 4.5 pK_a units compared with the same ion in free solution.[61] This propensity for an encapsulated, neutral species to show an increase in basicity was exploited by using the Ga_4L_6 cages to catalyze the acid-catalyzed hydrolysis of orthoformates and acetals in *alkaline* solution.[62] Under basic conditions, both classes of molecule are stable. Indeed, acetals are commonly employed as base-stable protecting groups in synthetic chemistry. However, in the presence of a catalytic amount of Ga_4L_6, both orthoformates and acetals were hydrolyzed in an alkaline medium (Figure 4.9).[62] As with the aza-Cope rearrangement, however, only smaller substrates were encapsulated by the complex; larger substrates were not hydrolyzed.

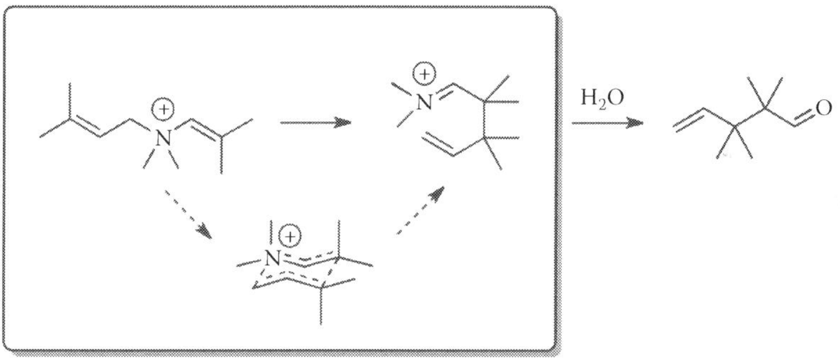

FIGURE 4.8 Example of the aza-Cope rearrangement catalyzed by M_4L_6 nanocages. Once the enammonium ion rearranges, it is quickly hydrolyzed by the aqueous environment.

(a)

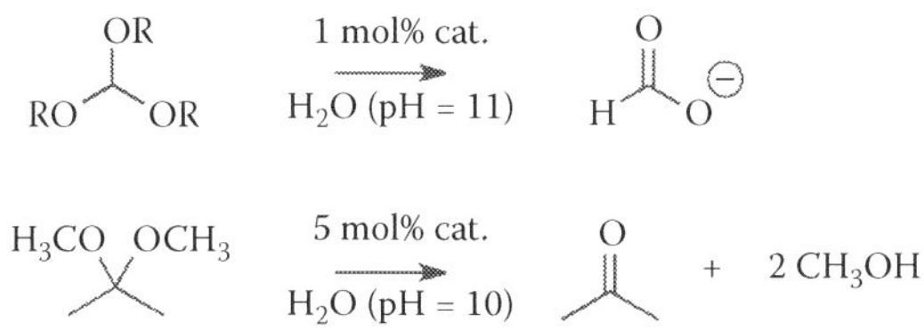

(b)

FIGURE 4.9 M_4L_6 complexes catalyze the acidic hydrolysis of orthoformates (a) and acetals (b) in alkaline solution.

4.3.2 Mimicking Enzyme Repair

Another particularly desirable trait of enzymes is their ability to repair when damaged. Catalysts, especially highly active ones, are prone to degradation during catalytic turnover. When a protein is damaged (e.g., an enzyme oxidizes one of its own amino acid residues "by accident"), the protein is dismantled and rebuilt or the damaged portion otherwise repaired.[63–65] Nonenzymatic catalysts are also prone to such damage, which reveals itself in low turnover numbers or the need to isolate and reactivate "spent" catalysts. Accordingly, there are efforts to develop self-healing catalysts that repair damage incurred during turnover and thus increase turnover and lengthen catalyst lifetime.

A catalyst that self-assembles by just mixing simpler precursors is an ideal candidate for a self-healing catalyst.[66,67] The strategy here is that, during turnover, a catalyst or catalytic site will decompose to its precursor pieces when damaged, and those pieces will then reassemble to regenerate the catalyst (Figure 4.10). Hill and Neumann have both reported on the use of polyoxometalates (POMs) as oxidation catalysts.[68,69] POMs are redox-active, discrete clusters of transition metal ions bridged by many oxygen atoms. The reported POM catalysts self-assemble from simpler metal oxides in solution and show great stability during catalysis, suggesting the ability to repair damage.

More recently (and dramatically), the Nocera group reported the synthesis of a catalyst composed of cobalt, oxygen, and phosphate that is both capable of water oxidation and shows the ability to repair itself.[70,71] The efficient oxidation of water is a highly desirable reaction used by nature to store solar energy in chemical bonds. A synthetic water oxidation catalyst, accordingly, offers a biomimetic method for storing energy and is an active area of research (see Chapter 7). The oxygen-evolving catalyst developed by Nocera et al. forms *in situ* from a solution of Co^{II} and phosphate.[71] The catalyst forms on the surface of electrodes when an electrical potential is applied, but corrodes back into solution (as Co^{II} and phosphate) when the circuit is turned off. This process could be directly monitored using radioactive ^{57}Co and ^{32}P, and the catalyst reforms anew when the potential is re-applied.[72] This self-repairing catalyst has since been incorporated into an artificial "leaf" that absorbs light and stores the energy in the chemical bonds of O_2 and H_2.[70]

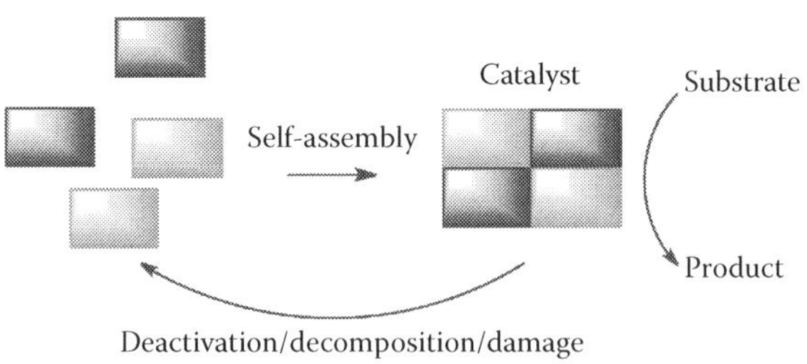

FIGURE 4.10 A catalyst that can properly self-assemble from simpler fragments under the conditions of catalytic turnover should be able to repair and reform if it is damaged during catalysis.

4.3.3 Mimicking the Active Site

Other biomimetic efforts focus on constructing synthetic models of the reactive functionalities found in an enzyme's active site. These models provide an understanding of the fundamental chemical reactions occurring during enzyme catalysis and are especially useful in cases where the enzyme can catalyze a reaction that remains a challenge (or even an impossibility) in the laboratory. In tandem with biochemical investigations of the enzyme itself, biomimetic models explain how various structural features of the active site affect the catalytic chemistry of the enzyme. Through such efforts, one learns how the reaction can be controlled to develop a practical, synthetic catalyst that mimics the enzyme's function. The final, practical product may look nothing like nature's version, but the inspiration and the lessons learned from biology make that final product possible. As Breslow has put it, "In biomimetic chemistry, we ... take inspiration, but not blueprints, from natural chemistry."[73]

Using synthetic organic and inorganic techniques, the structure of the model can be systematically modified to recreate the activity of the enzyme and understand those features that "make it work." This is especially useful for studying the active sites of metalloenzymes in which not only the chemical structure of the active site but also factors such as the metal's oxidation state and coordination can have dramatic importance to reactivity. Studies of synthetic models can work toward a variety of ends, including:

- Elucidate the mechanism of an enzymatic reaction
- Stabilize, isolate, and study known or proposed reactive intermediates of the enzymatic or model compound's catalytic cycle
- Investigate how covalent modifications to the model affect reactivity
- Reveal how the reaction medium (e.g., aqueous, acidic, nonpolar) affect the model's activity
- Explore the relationship between model reactivity and some physical or chemical property (e.g., bond length and reduction potential)
- Develop a practical catalyst that mimics the desired enzymatic activity

A particularly active area for model studies is the development of mimics of cytochrome P450.[34] As described previously, P450 enzymes catalyze the insertion of an oxygen atom from O_2 into a C–H bond. The active site of P450 contains an iron porphyrin macrocycle known as heme (Figure 4.11). Heme is ubiquitous in biochemistry, playing roles in O_2 transport (hemoglobin) and electron transfer (cytochrome *c*) in addition to catalysis. A major focus of heme model studies has been to understand how subtle changes to the iron and porphyrin environment can lead to such dramatic differences in activity (e.g., what makes heme sometimes an O_2 carrier and sometimes an oxidation catalyst?).

The consensus mechanism for P450 catalysis is understood in much detail,[35] although it is beyond the scope of this discussion. In brief, O_2 (with the assistance of two electrons and protons) oxidizes the resting iron(III) state of the heme to an oxoiron(IV) porphyrin cation radical that has been historically named "Compound I"

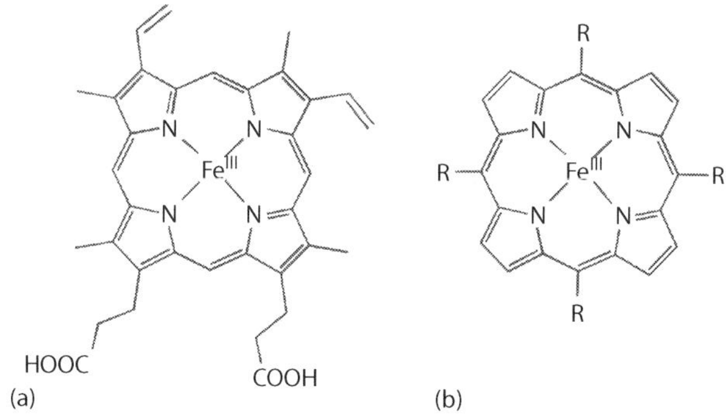

FIGURE 4.11 (a) Structure of heme (protoporphyrin IX) and (b) a synthetic, *meso*-substituted iron porphyrin.

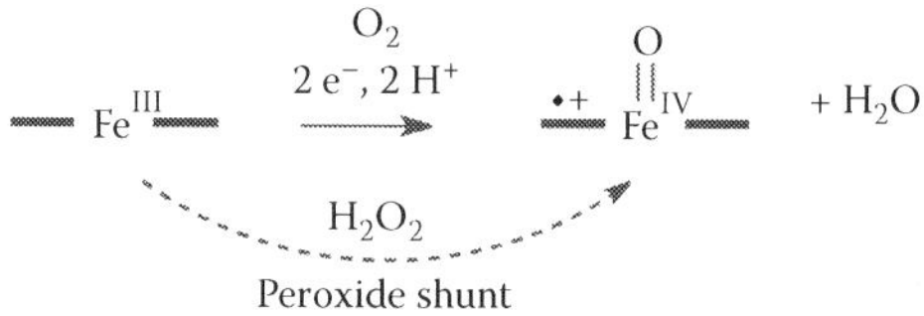

FIGURE 4.12 Generation of Compound I from iron(III) heme in cytochrome P450. (The porphyrin macrocycle is abbreviated here as two bolded lines, as if we were looking side-on at the ring.)

(Figure 4.12).* Compound I carries out the hydroxylation reaction characteristic of P450 enzymes, producing an alcohol product and regenerating the starting iron(III). The iron(III) heme can also be oxidized to Compound I using hydrogen peroxide (known as the "peroxide shunt" pathway).

Compound I hydroxylates a substrate by what has become known as "oxygen rebound" (Figure 4.13). First, the oxoiron abstracts a hydrogen atom from a C–H bond, generating a carbon-centered radical and an iron-bound hydroxyl. The oxygen of the hydroxyl then "rebounds" to the carboradical to produce the product alcohol. "Oxygen rebound" was first proposed in 1976 based on investigations of oxygen-atom transfer by a synthetic iron porphyrin model system.[74]

The mechanism of P450 catalysis has been elucidated through the use of synthetic model iron porphyrins (Figure 4.11b) that have allowed the generation, isolation, and characterization of purported intermediates. Long before the structure of Compound I was confirmed in the P450 enzyme, a Compound I analogue was generated and characterized in a synthetic porphyrin complex and shown to hydroxylate cyclohexane.[74] Moreover, model studies using synthetic metalloporphyrins (metal porphyrin

* Although Compound I is formally an oxoiron(V) species, one of the oxidizing equivalents resides on the porphyrin macrocycle and not the metal center.

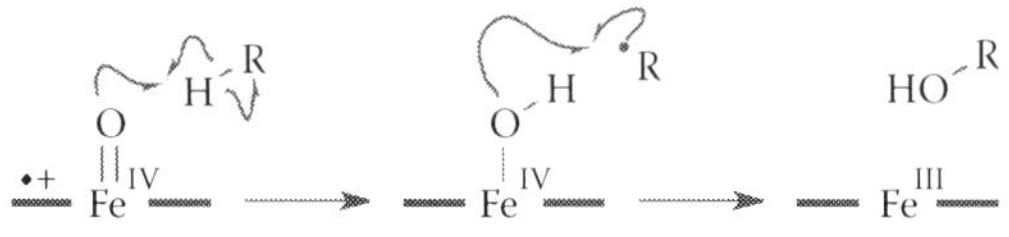

FIGURE 4.13 Mechanism of oxygen rebound. (The porphyrin macrocycle is abbreviated here as two bolded lines, as if we were looking side-on at the ring.)

complexes) have revealed much about the stereoelectronic factors that mediate the activity of high-valent oxoiron compounds to oxidize substrates. For example, modification of the porphyrin ligand with electron withdrawing substituents dramatically increase the reactivity of Compound I models.[75]

Although nature chose iron as the metal for heme, model studies are not so limited and have included many other metals. Notably, manganese porphyrins have also been shown to catalyze the oxidation of organic substrates using oxidants such as hydrogen peroxide, peroxyacids, and sodium hypochlorite. Generally, the manganese porphyrins have outperformed the iron versions as oxidation catalysts.[34] The manganese version of iron's Compound I is an oxomanganese(V) species. Interestingly, although electron withdrawing substituents increase the reactivity of Compound I, the same substituents stabilize oxomanganese(V) porphyrins.[76,77] Additionally, pH greatly modulates the reactivity of the oxomanganese(V) porphyrin. The oxomanganese(V) species exists in actuality as a *trans*-dioxomanganese(V) species that requires a proton to react (Figure 4.14).[78] At high pH, the dioxomanganese(V) cannot receive this proton and catalysis is interrupted, allowing one to conceivably control the reaction simply through pH.

The majority of metalloporphyrin studies have focused on understanding the mechanism of iron-catalyzed, controlled oxidation as well as those structural and other factors that most influence catalysis.[34] In that regard, metalloporphyrins have greatly succeeded. The knowledge gained has not only enriched our understanding of P450 itself but also informed the development of many practical oxidation catalysts using abundant, first-row transition metals. Significant efforts have been dedicated toward developing a more diverse class of nonheme (i.e., nonporphyrin) iron and manganese model systems.[79] For example, Que and coworkers demonstrated a stereospecific alkane hydroxylation using a nonheme iron model system iron(II) TPA$(CH_3CN)_2[ClO_4]_2$ (Figure 4.15) and hydrogen peroxide.[80] Additionally, several ligand systems have allowed the in-depth characterization of important intermediates

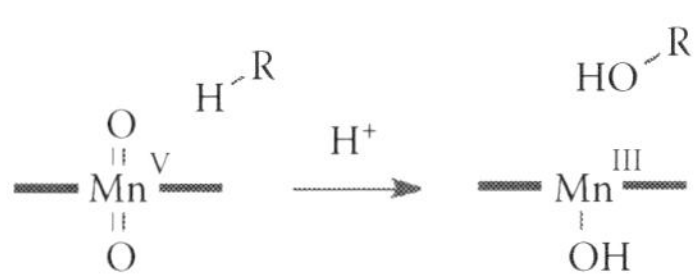

FIGURE 4.14 *Trans*-dioxomanganese(V) porphyrins, the manganese analogue of Compound I requires a proton (H^+) to transfer an oxygen atom to a substrate. (The porphyrin macrocycle is abbreviated here as two bolded lines, as if we were looking side-on at the ring.)

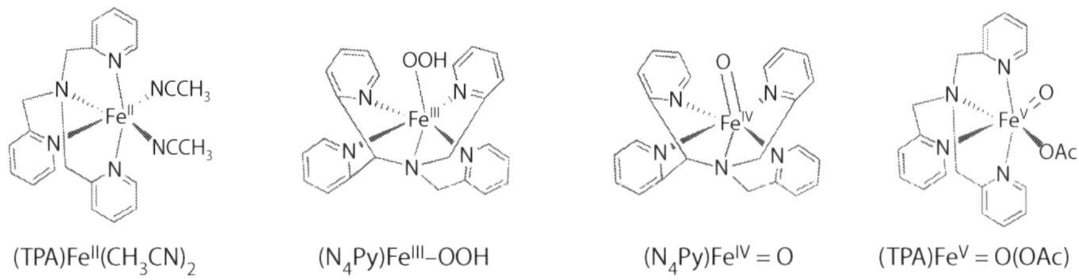

FIGURE 4.15 Spectroscopically characterized nonheme iron intermediates. (Adapted from Kim, C. et al., *J Am Chem Soc*, 119 (25), 5964–5965, 1997; Ray, K. et al., *J. Am. Chem. Soc.*, 136 (40), 13942–13958, 2014.)

in iron-catalyzed oxidations (e.g., hydroperoxoiron(II)-OOH, oxoiron(IV), and oxoiron(V); Figure 4.15), further aiding mechanistic studies of both catalyst and enzyme.[81]

Biomimetic model systems are not limited merely to C–H hydroxylation. Models are studied for a variety of other enzymes including (but by no means limited to) Photosystem II,[82] hydrogenase,[83] and methane monoxygenase,[84] and such models have aided in our understanding of how those enzymes operate. Biomimetic models have informed the development of a number of practical catalysts or processes. Representative examples of such inspired catalysts are the focus of the remainder of this chapter.

4.4 PRACTICAL APPLICATIONS OF BIOMIMETIC CHEMISTRY

4.4.1 Total Synthesis

Incorporating chiral carbon centers into complex organic molecules is necessary in the synthesis of many pharmaceutical agents.[85] One strategy builds upon the inherent reactivity of epoxides, which can be ring-opened stereoselectively and regioselectively by a number of nucleophiles to quickly add complexity to a molecular structure. Therefore, the development of asymmetric catalysts suitable for the production of enantiomerically pure epoxides, important building blocks for complex synthesis, represents an important goal.[86]

Jacobsen and coworkers met this need by developing a manganese salen complex that catalytically epoxidizes alkenes (Figure 4.16). Using hydrogen peroxide or hypochlorite as oxidants, "Jacobsen's catalyst" performs chiral epoxidation of olefins with high stereoselectivity.[87] The potential of the catalyst was demonstrated in the asymmetric synthesis of the side chain of Taxol in 1992 (Figure 4.17).[88] Taxol, an important anticancer natural product, is on the World Health Organization's List of Essential Medicines.[89] (Due to its intensive, low-yielding syntheses, Taxol is now industrially prepared in high yield through a fermentative process developed by Bristol-Meyers Squibb that eliminates chemical waste and saves energy. This technology was awarded a Presidential Green Chemistry Challenge Award in 2004.[90])

Many diverse modifications of Jacobsen's catalyst have been undertaken to design increasingly more complex catalysts;[91] however, it is the simplicity of the original

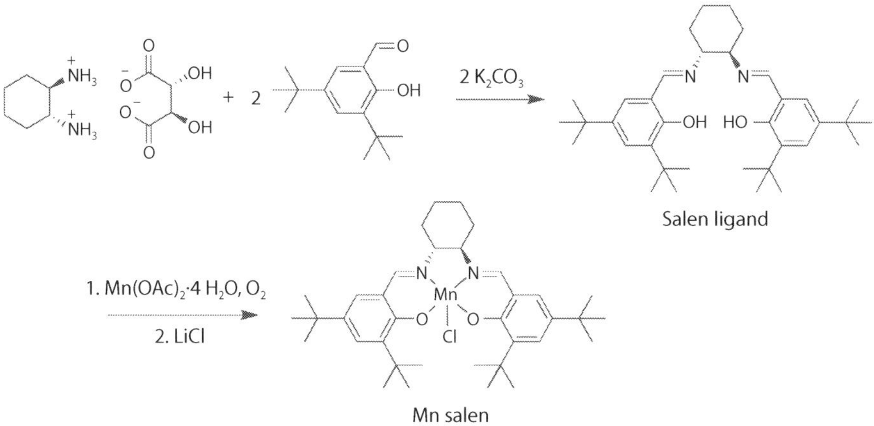

FIGURE 4.16 Synthesis of Jacobsen's catalyst (*R,R*) "*N,N*"-bis(3,5-di-*tert*-butylsalicylidene-1,2-cyclohexanediaminomanganese(III) chloride). (Adapted from Deng, L., and Jacobsen, E.N. *J Org Chem*, 57 (15), 4320–4323, 1992.)

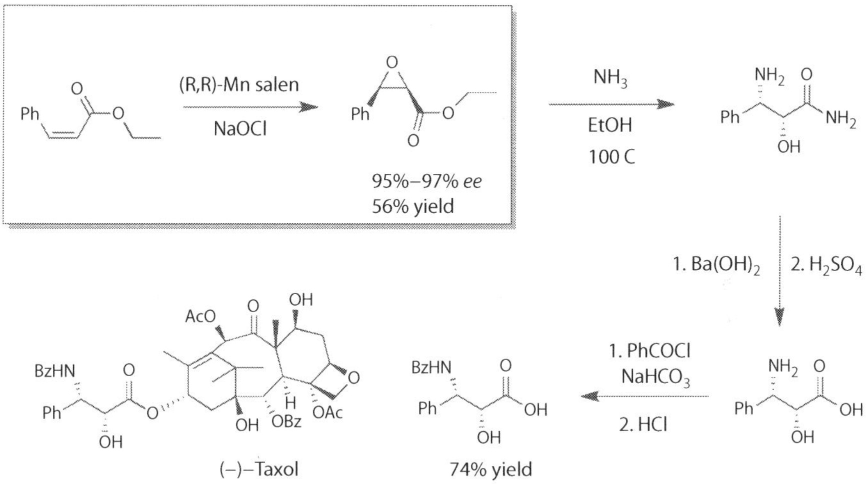

FIGURE 4.17 Synthesis of the side chain of Taxol. (Adapted from Deng, L., and Jacobsen, E.N. *J Org Chem*, 57 (15), 4320–4323, 1992.)

ligand synthesis that defines its utility. "Jacobsen's catalyst" can be prepared from simple components, minimizing the time, energy, and resources necessary to formulate it.[88] (The synthesis and application of this catalyst has become a common sequence of undergraduate laboratory experiments, demonstrating its ease of preparation and use.)[92] Furthermore, the relatively simple synthesis and structure demonstrate that an impactful catalyst need not be complex. The use of hydrogen peroxide as an oxidant is also particularly attractive, as the by-product of its use as an oxidant is water. Built from simple components and employing clean reagents for catalysis, Jacobsen's catalyst is capable of generating high-value products meeting a great need while minimizing costs and environmental impacts.

4.4.2 Late Stage Modifications

Natural products, organic compounds isolated from biological sources such as plants or microbes, often show great promise as pharmaceuticals due to their high bioactivity, but they often have characteristics that make them poor candidates for development as a drug: complex syntheses, low solubility, poor pharmacological properties, and low availability from natural sources.[93] One approach aims to generate analogues of the desired natural product and relies on the ability to make "late stage" modifications to complex molecules (i.e., specifically modifying a complex compound's structure near or at the end of its synthesis). Advancing a promising compound, from an early "hit" in a drug discovery program through the stages of drug development to market, nearly always requires extensive modification of the original structure.[93]

Approximately 80% of commercial drugs derived from natural products require synthetic modification and improvements to overcome issues with the availability of the compound or pharmacological properties such as solubility, stability, or toxicology.[93] Optimization of a natural product drug candidate is often extremely difficult due to the necessity to directly modify the core structure of the natural product without unwanted modifications to the rest of the molecular architecture. A total synthesis of the modified candidate from scratch is an option, but natural products often require extensive and low-yielding syntheses that prohibit this strategy. (The shortest reported total synthesis of Taxol, discussed in the previous section, requires 36 synthetic steps.)[94] Making derivatives "from scratch" for a molecule as complex as Taxol can be economically prohibitive; thus, a more desirable strategy begins with the isolated natural product candidate and directly and selectivity modifies its structure.

Late-stage modification often requires the replacement an unreactive C–H bond with some other functionality. To quickly and efficiently modify a complex molecule in a controlled fashion would require one to be able to view the C–H bond as "just another reactive functional group." Using the lessons learned from biomimetic models of C–H activation enzymes, direct, controlled, and specific functionalizations of a molecule can be realized. Such advances would have a remarkable effect on the types of compounds available to study as drugs by keeping promising natural products in the development pipeline.

Motivated by nature's ability to control C–H functionalization on complex molecules, chemists have begun to apply lessons from biomimetic study in their synthetic methodology.[95] For example, Chen and White demonstrated an elegant and useful example of nonheme iron catalysis for late-stage modifications.[96] Using the bulky, electrophilic catalyst Fe(S,S-PDP) (Figure 4.18b), hydrogen peroxide as an oxidant, and acetic acid as the only additive, they demonstrated selective C–H oxidation based on subtle differences between multiple tertiary C–H bonds. This is shown by the hydroxylation of artemisinin, an important antimalarial compound. The Fe(S,S,-PDP)-catalyzed hydroxylation of artemisinin proceeded with selectivity for the C6 tertiary C–H position in the presence of five tertiary C–H bonds (Figure 4.18).

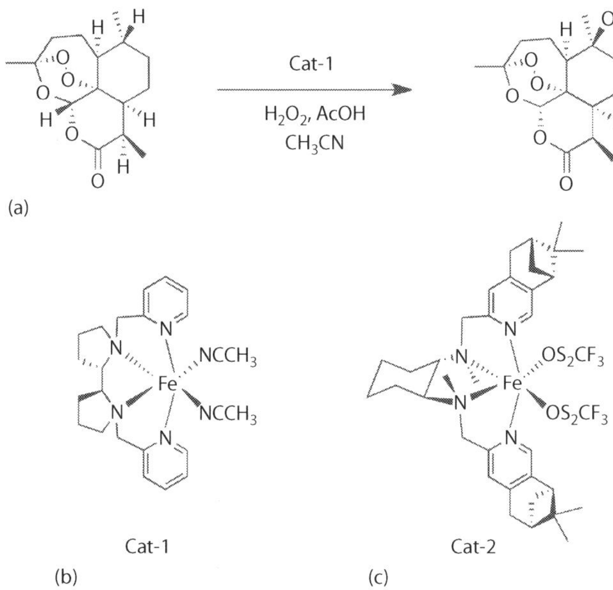

FIGURE 4.18 (a) Selective oxidation of artemisinin using an $Fe(CH_3CN)_2$(S,S-PDP) catalyst. (b) $Fe(CH_3CN)_2$(S,S-PDP). (c) $Fe(CF_3SO_3)_2$((S,S,R)-MCPP).

Because a tertiary C–H bond is particularly weak among C–H bonds, it is unsurprising that a tertiary position was oxidized; however, it is remarkable that the catalyst would select for one specific tertiary position over the others. This ability to so selectivity control the oxidation of a complex natural product is an important step for widespread late-state diversification.

Further demonstrations have shown selectivity with the Fe(S,S-PDP) catalyst between secondary C–H bonds for a diverse class of substrates.[97] Product conversions with the Fe(S,S-PDP) ligand are relatively high for several complex substrates. A more complex catalyst prepared by Costas (Figure 4.18c) had increased stability allowing lower catalyst loading.[98] From the perspective of the pursuit of sustainable catalysis, then, balancing the overall costs of increasingly complex catalysts with the benefits of higher yield and activity will need to be considered on a case-by-case basis.

Directing and controlling for the site of modification can be challenging. Sometimes, a neighboring functional group can assist in directing reactivity toward a specific C–H bond. For example, a collaboration between Phil Baran and Jin-Quan Yu's groups at Scripps recently demonstrated the ability to generate a library of compounds based on the natural product (+)-Hongoquercin A.[99] The ability to generate a library of analogues allows one to rapidly screen for and optimize some desired property. Here, multiple analogues are produced through substitution of the C–H para to the carboxylic acid by a suite of reactions (Figure 4.19). The carboxylic acid moiety on (+)-Hongoquercin A helps direct C–H activation to a specific position. As each natural product's structure is unique, the need for a specific directing group often hinders

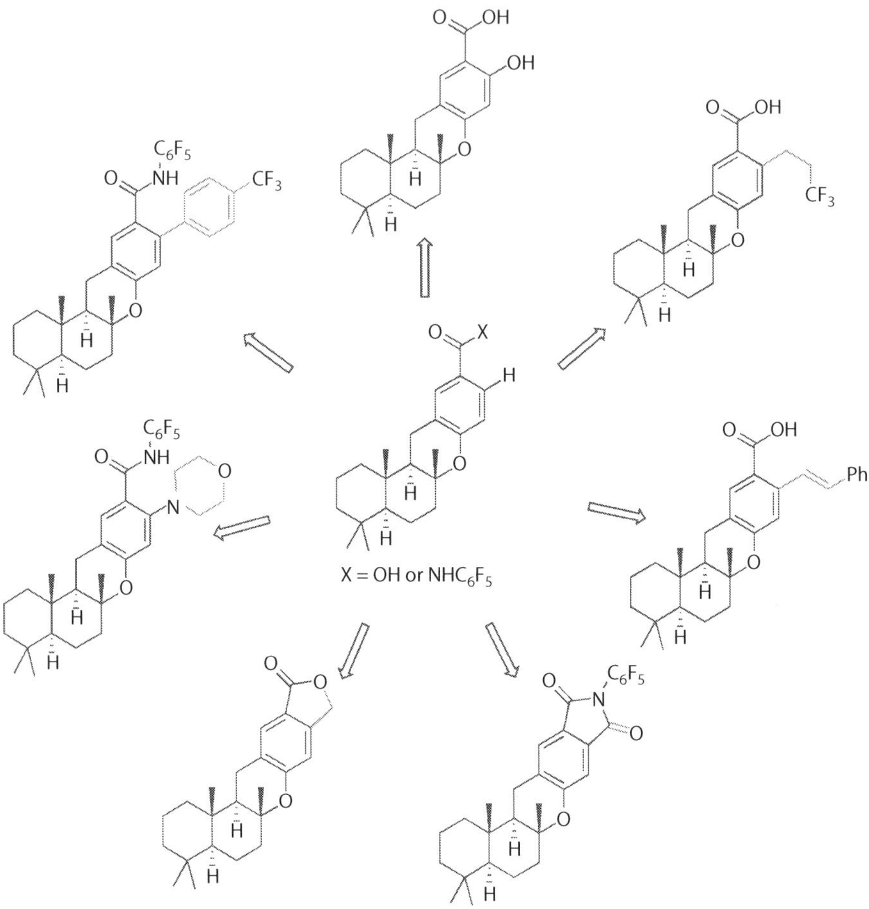

FIGURE 4.19 Examples of using C–H functionalization to generate a compound series based on a natural product. (Adapted from Rosen, B.R. et al., *Angew Chem Int Ed*, 52 (28), 7317–7320, 2013.)

a more general application. Thus, improved techniques for the direct modification of complex natural products remains an important and ongoing challenge.[100]

4.4.3 Biomimetic Oxidation for Metabolite Production

Understanding the metabolism of potential drug candidates in the human body is an extremely important aspect of drug development.[101] The first route for the elimination of drugs from the body are the enzymatic oxidation reactions undertaken by cytochrome P450 enzymes in the liver.[37] When determining drug safety and toxicity, it is imperative to understand and identify *all* of the potential metabolites generated *in vivo*. Metabolites may have higher activity than the parent drug (necessitating intellectual property protection), increased toxicity, or extended half-life, distribution, or excretion profiles.[102] For example, the toxicity of acetaminophen is caused not by the drug itself but a toxic metabolite generated during drug metabolism. Toxicity issues account for approximately 20% of late stage drug trial failures,[103] whereas recalling an approved drug from market due to toxicity issues can cost millions or even billions of dollars in lost revenue and damages.[104,105] Clearly, there are aspects of drug development that are not socially or economically sustainable.

The FDA has issued the metabolites in safety testing guidelines in 2008 outlining safety tests regarding metabolites in drug development.[106] The recommendations include independent safety testing for all major metabolites found in human plasma at greater than 10% of parent exposure as well as all metabolites that are unique in humans not expressed in standard toxicological models. Compliance requires the identification of metabolites of drug metabolism as well as the synthesis of sufficient quantities for testing, both of which can be extremely difficult. Metabolites found *in vivo* are often unstable or reactive and produced in small quantities, making it very difficult to isolate enough material to determine structure.[107] Due to the extreme importance of metabolite safety testing, many approaches to identification and production of metabolites have been established.[108] These include using liver slices, microsomes, recombinant enzymes or hepatocytes in complex cell cultures, or the metabolite(s) may be directly isolated from plasma or urine. Alternatively, a proposed structure can be prepared by total synthesis. Each of these approaches can be time-consuming and economically prohibitive, and might still not isolate or produce enough of the desired compound (or the correct compound) to be useful.

An increasingly common method for metabolite production involves the use of biomimetic catalysis.[108] The extensive research undertaken to develop model systems for C–H oxidation, specifically P450-mimicking metalloporphyrins, makes model systems a logical choice for studying drug metabolisms.[109–111] Metalloporporphyrins catalyze a broad range of cytochrome P450 reactions such as hydroxylation (aromatic and aliphatic), epoxidation, N-oxidation, and S-oxidation (Figure 4.20).[35,76,112–115] Because the metalloporphyrin systems mimic different aspects of P450 metabolism, chemical screens ("kits") of these various catalyst systems allow one to rapidly produce a library of potential metabolites. From such screens, one can quickly optimize product conversion and selectivity and find a reaction condition for the metabolite production of interest *directly from the parent drug compound in a one-step synthesis*.[108] Compared with the time required for optimizing (e.g., mutating, engineering) an enzyme to improve biocatalysis, biomimetic oxidation can show significantly shorter timelines to access significant quantities of a metabolite.

Although unlikely to ever directly replace the need for microsomes or whole cell studies to identify human metabolites, biomimetic screens offer an alternative approach to metabolite production with several advantages compared with the biological options. Higher concentrations of substrates can be used in biomimetic metabolite production than enzymes or cells can tolerate, which facilitates higher amounts of product. Moreover, the screens preclude the need to remove buffers and proteins from the reaction mixture, leading to enhanced product yields. The smaller solvent volumes necessarily make purification easier and equals less waste and energy use. Finally, access to metabolites at a large scale early in the drug development process can assist in identifying toxicology failures well before the clinic, saving money and, more importantly, lives.

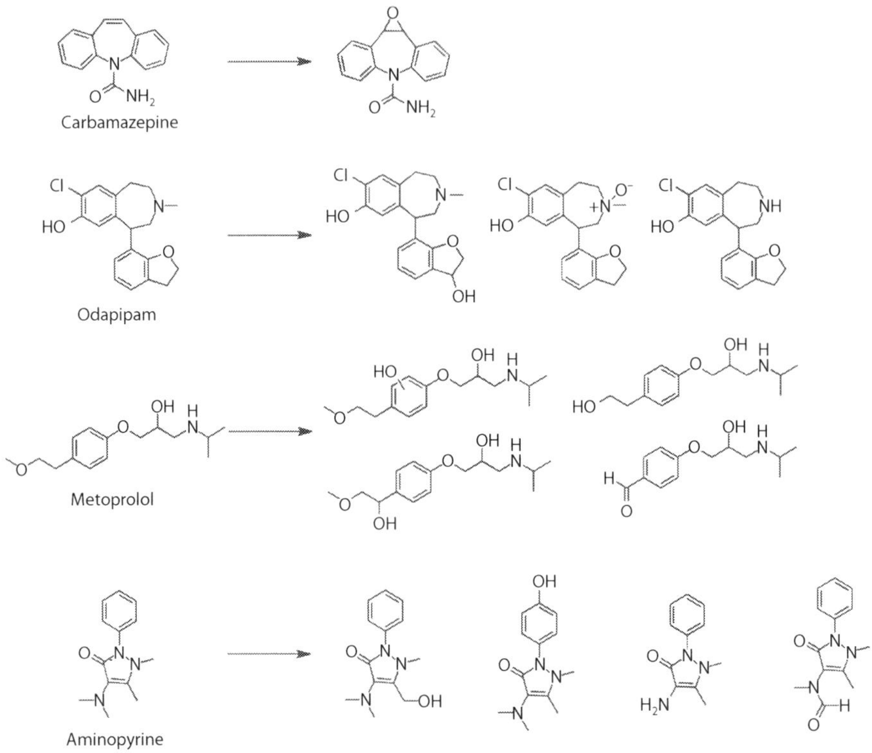

FIGURE 4.20 Examples of selective oxidations of known pharmaceuticals catalyzed by biomimetic metalloporphyrins. (Adapted from Guengerich, F.P. *AAPS J*, 8 (1), E101–E111, 2006; Groves, J.T. et al., *J Am Chem Soc*, 119 (27), 6269–6273, 1997; Zhang, K.Y.E. et al., *Antimicrob Agents Chemother*, 45 (4), 1086–1093, 2001; Barret, R. et al., *Pharmazie*, 42 (2), 132–132, 1987.)

4.4.4 Fluorination

The ability of the cytochrome P450 enzymes to oxidize a drug, although a source of much chemical inspiration, can also be the downfall of the potential therapeutic promise. The human body uses cytochrome P450s to metabolize a drug that it perceives as toxin, facilitating its excretion from the body. However, to be effective, the drug needs to remain in the body long enough to achieve its desired therapeutic effect. This trade-off is one of the most important aspects of drug development.[116] What does one do about a drug that shows incredible promise for treating a target condition but is quickly oxidized (metabolized) by the body and deactivated before it has a chance to act?

Medicinal chemists often look to modify a drug molecule to increase its stability toward P450s and generally increase the lifetime of the compound in the body while trying not to lose the therapeutic activity. Late-stage fluorination, particularly at sites of a molecule where metabolism occurs, is particularly of value. The C–F bond is stronger than the C–H bond, which dramatically increases compound stability.[117] In fact, fluorination is an area where chemists have innovated ahead of biology. Although approximately 20% to 25% of known drugs contain at least one fluorine atom, including blockbusters such as Prozac, Celebrex, and Lipitor,[118] only five confirmed fluorinated natural products are known to date.[119] Of course, because the C–F bond is so stable in the body, fluorinated compounds are also persistent in the environment and do not degrade readily.[120] Accordingly, another trade-off will eventually need to be considered between the economic and social benefits of a more effective drug and the environmental impacts of a persistent pollutant when the drug is excreted. For the time being, however, fluorination is a practiced strategy that can benefit from improvements with regard to safety, time, and cost.

Fluorination chemistry has, until recently, been its own "niche" field, often requiring dangerous reaction conditions and relying on harsh reagents (such as F_2 gas or hydrofluoric acid) to achieve unselective reactions.[121] However, the development of air- and moisture-stable "benchtop" fluorination reagents have safely allowed the expansion of fluorination chemistry into the realm of organometallic catalysis and more traditional chemical settings.[122] Recent examples demonstrate the utility of organometallics for fluorination, such as Sanford and coworkers' use of palladium catalyzed nucleophilic fluorination using silver fluoride.[123]

Inspired by P450 oxygen rebound, Groves and coworkers reported a C–H fluorination reaction using a biomimetic, manganese porphyrin catalyst.[124] Using a manganese(III) porphyrin catalyst, iodosylbenzene as an oxidant, and silver fluoride as the fluorine source, the team selectively fluorinated simple alkanes, terpenoids, and steroids at generally inaccessible sites. The proposed mechanism, similar to P450 oxygen rebound, suggests that an oxomanganese(V) abstracts a hydrogen atom from a C–H bond. Unlike P450 and other models, however, a manganese-bound fluoride here rebounds to the carboradical (Figure 4.21) to generate the fluorinated product. Indeed, a classic biomimetic system like the manganese porphyrins can still guide the development of a practical application even after almost 40 years.

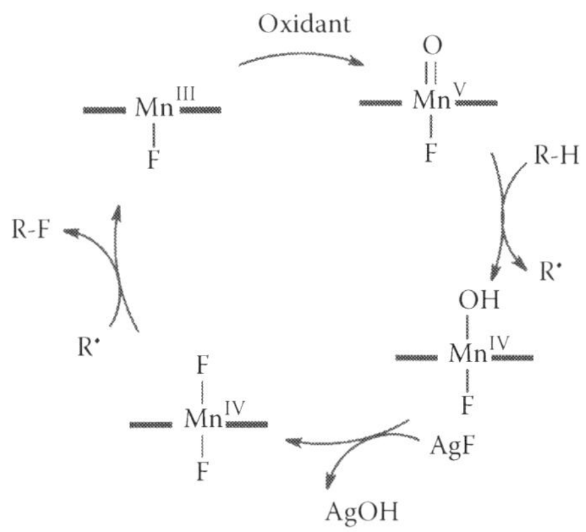

FIGURE 4.21 Proposed mechanism of C–H fluorination by a biomimetic manganese porphyrin catalyst. The porphyrin macrocycle is abbreviated here as two bolded lines, as if we were looking side-on at the ring. (Adapted from Liu, W. et al., *Science*, 337 (6100), 1322–1332, 2012.)

The extension of this system to others ligands, including a simple salen, demonstrates the robustness of the reaction, and has potential not only for late stage fluorination but also the fast and efficient incorporation of radioactive ^{19}F atoms into compounds. Fluorine-19 is used as a positron source in PET diagnostic imaging, and the isotope is introduced as part of a fluorinated compound.[125] As ^{19}F has a half-life of less than 20 min, a fast and easy method for introducing radioactive fluorine into a complex molecule is necessary for advanced PET studies. The reaction conditions (necessitating the chlorinated solvent dichloromethane) are nonideal from a green chemistry standpoint, but the social benefits may outweigh the environmental benefits in the short-term until a "greener" version of this process can be developed. Advances in fluorinated techniques will only serve to increase the number of new drugs containing fluorine, including more drugs derived from natural products. As such, there is a great need to consider sustainability as methods suitable for industrial scale are developed.

4.4.5 Water Purification

Oxidation reactions are useful not only for the controlled and selective functionalization of molecules as described in the examples above. Oxidation reactions can also be used as a less-selective way to degrade or deactivate pollutants and contaminants. In perhaps the most direct overlap between biomimicry and green chemistry, the tetraamido macrocyclic ligand (TAML) peroxide activators serve the purpose of providing robust, broad, and "green" oxidations.

TAML catalysts (Figure 4.22) are iron-containing catalysts that activate hydrogen peroxide in a fashion similar to the cytochromes P450 or peroxidase enzymes. Presumably, hydrogen peroxide oxidizes the Fe-TAML to generate some species, akin to a high-valent iron-oxo as in the heme enzymes, that oxidizes organic matter.[126] Fe-TAML complexes are broad-spectrum oxidation catalysts, having activity toward a variety of organic substances at neutral pH (although they are also active in highly alkaline environments).[127] Notably, TAML catalysts were developed in an iterative sequence that maximized the number of turnovers by "designing out" portions of the ligand prone to decomposition.[128] In practice, TAML catalysts can have

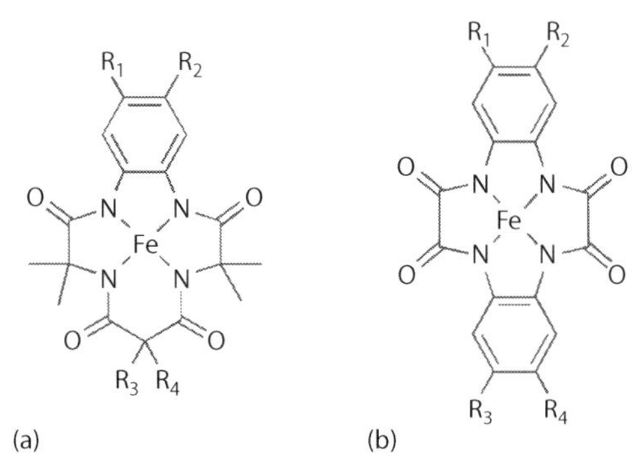

FIGURE 4.22 First- (a) and second- (b) generation TAML catalysts. (From Ellis, W.C. et al., *J Am Chem Soc*, 132 (28), 9774–9781, 2010.)

turnover frequencies of 10,000 hr^{-1}.[128] Unlike many of the biomimetic catalysts discussed above, the intended purpose of the Fe-TAML oxidation catalysts is to *nonselectively* oxidize organic matter and degrade it as much as possible.

Fe-TAML complexes are applied as homogeneous oxidation catalysts for paper (pulp) bleaching, stain removal, and water purification. In paper bleaching, Fe-TAML/hydrogen peroxide systems replace traditional chlorine-based methods. Whereas chlorine bleaching generates chlorinated organics and salts as by-products,[47] the TAML/hydrogen peroxide system simply produces water. In water purification, Fe-TAML/hydrogen peroxide systems have been successfully applied to oxidize and degrade persistent organic pollutants including pharmaceuticals,[129] endocrine-disrupting estrogens,[130] pesticides,[131] dyes,[132] and explosives.[133] Although homogeneous catalysts are typically problematic because they must be removed upon the completion of catalysis, Fe-TAML complexes were developed not only to maximize their catalytic efficiency but also to minimize their toxicity[134] (precluding their need to be reclaimed in many circumstances). The TAML technology was honored with a Presidential Green Chemistry Challenge Award in 1999.[135]

4.5 THE FUTURE OF BIOMIMICRY

What does the future hold for biomimetic catalysis? More than 40 years since Breslow first tried to mimic an enzyme, an entire research field has developed around "doing what nature does." Originally, the goal was to simply build and design a better catalyst and was fueled by curiosity. Over time, basic, fundamental research to understand the reactivity of an iron enzyme (cytochrome P450) led to simple model systems that illuminated how the enzyme worked. Biomimicry has informed other fields, providing unforeseen insights to fundamental chemical reactivity. Now, biomimetic catalysts have real applicability to solve needed challenges in chemistry.

A new challenge has emerged, however. Sustainability calls us to build a better catalyst as well, but "better" does not merely mean "faster" anymore. The new standard for success includes not only accomplishing the desired reaction but also doing so while simultaneously addressing the concerns of the people, the planet, and profits. Nature provided us with a wealth of examples of ideal catalysts in the form of enzymes. Not only do enzymes tantalize us with reactions we wish we could better control (e.g., late state C–H functionalization), but enzymes show us that we can do important chemistry with cheap and abundant materials, under ambient conditions, and with the production of little or no waste. The future of biomimetic catalysis will do more than teach us how an enzyme works. We now need to focus on taking those lessons and putting them into practice.

REFERENCES

1. Brocchieri, L.; Karlin, S., *Nucleic Acids Res.* 2005, *33* (10), 3390–3400.
2. International Union of Biochemistry and Molecular Biology. Available at http://www.iubmb.org/ (accessed November 14, 2014).
3. Nestl, B. M.; Hammer, S. C.; Nebel, B. A.; Hauer, B., *Angew. Chem., Int. Ed.* 2014, *53* (12), 3070–3095.
4. Davis, B. G.; Boyer, V., *Nat. Prod. Rep.* 2001, *18* (6), 618–640.

5. Wolfenden, R.; Snider, M. J., *Acc. Chem. Res.* 2001, *34* (12), 938–945.
6. Radzicka, A.; Wolfenden, R., *Science* 1995, *267* (5194), 90–93.
7. Hall, A.; Knowles, J. R., *Biochemistry* 1975, *14* (19), 4348–4353.
8. Putman, S. J.; Coulson, A. F.; Farley, I. R.; Riddleston, B.; Knowles, J. R., *Biochem. J.* 1972, *129* (2), 301–310.
9. Blacklow, S. C.; Raines, R. T.; Lim, W. A.; Zamore, P. D.; Knowles, J. R., *Biochemistry* 1988, *27* (4), 1158–1167.
10. Knowles, J. R., *Nature* 1991, *350* (6314), 121–124.
11. Benkovic, S. J.; Hammes-Schiffer, S., *Science* 2003, *301* (5637), 1196–1202.
12. Bugg, T., *Introduction to Enzyme and Coenzyme Chemistry*, 2nd ed.; Blackwell Pub.: Oxford, UK; Malden, MA, 2004.
13. Bruice, T. C.; Benkovic, S. J., *Biochemistry* 2000, *39* (21), 6267–6274.
14. Hedstrom, L. Enzyme specificity and selectivity. In *eLS*. John Wiley & Sons Ltd.: Chichester, (Feb 2010). Available at http://www.els.net [doi: 10.1002/9780470015902.a0000716.pub2].
15. Fischer, E., *Ber. Dtsch. Chem. Ges.* 1894, *27*, 3189.
16. Staunton, J.; Wilkinson, B., *Chem. Rev.* 1997, *97* (7), 2611–2629.
17. Madshus, I. H., *Biochem. J.* 1988, *250* (1), 1–8.
18. Feldhamer, G. A., *Mammalogy: Adaptation, Diversity, and Ecology*. WCB/McGraw-Hill: Boston, 1999.
19. Elleuche, S.; Schroder, C.; Sahm, K.; Antranikian, G., *Curr. Opin. Biotechnol.* 2014, *29C*, 116–123.
20. Demirjian, D. C.; Moris-Varas, F.; Cassidy, C. S., *Curr. Opin. Chem. Biol.* 2001, *5* (2), 144–151.
21. Gerday, C.; Aittaleb, M.; Bentahir, M.; Chessa, J. P.; Claverie, P.; Collins, T.; D'Amico, S.; et al. *Trends Biotechnol.* 2000, *18* (3), 103–107.
22. Stiefel, E. I., Bioinorganic chemistry and the biogeochemical cycles. In *Biological Inorganic Chemistry: Structure and Reactivity*, Bertini, I.; Gray, H. B.; Stiefel, E. I.; Valentine, J. S., Eds. University Science Books: Sausalito, CA, 2007; pp. 7–30.
23. Voet, D.; Voet, J. G., *Biochemistry*, 3rd ed.; John Wiley & Sons: New York, 2004.
24. Bertini, I.; Gray, H. B.; Stiefel, E. I.; Valentine, J. S., *Biological Inorganic Chemistry: Structure and Reactivity*. University Science Books: Sausalito, CA, 2007.
25. Chondrogianni, N.; Petropoulos, I.; Grimm, S.; Georgila, K.; Catalgol, B.; Friguet, B.; Grune, T.; Gonos, E. S., *Mol. Aspects Med.* 2014, *35*, 1–71.
26. Bommarius, A. S.; Riebel, B. R., *Biocatalysis*. Wiley-VCH: Weinheim; Cambridge, 2004.
27. Quin, M. B.; Schmidt-Dannert, C., *ACS Catal.* 2011, *1* (9), 1017–1021.
28. Schoemaker, H. E.; Mink, D.; Wubbolts, M. G., *Science* 2003, *299* (5613), 1694–1697.
29. Benyus, J. M., *Biomimicry: Innovation Inspired by Nature*, 1st ed.; Morrow: New York, 1997.
30. Breslow, R., *Chem. Soc. Rev.* 1972, *1* (4), 553–580.
31. Breslow, R., *Acc. Chem. Res.* 1980, *13* (6), 170–177.
32. Breslow, R., *Acc. Chem. Res.* 1995, *28* (3), 146–153.
33. Sono, M.; Roach, M. P.; Coulter, E. D.; Dawson, J. H., *Chem. Rev.* 1996, *96* (7), 2841–2887.
34. Groves, J. T., Models and mechanisms of cytochrome P450 action. In *Cytochrome P450: Structure, Mechanism, and Biochemistry*, 3rd ed.; Ortiz de Montellano, P. R., Ed. Kluwer Academic/Plenum Publishers: New York, 2005; pp. 1–44.
35. Ortiz de Montellano, P. R., *Chem. Rev.* 2010, *110* (2), 932–948.
36. Guengerich, F. P., *AAPS J.* 2006, *8* (1), E101–E111.
37. Denisov, I. G.; Makris, T. M.; Sligar, S. G.; Schlichting, I., *Chem. Rev.* 2005, *105* (6), 2253–2277.
38. Klingenberg, M., *Arch. Biochem. Biophys.* 1958, *75* (2), 376–386.

39. Omura, T.; Sato, R., *J. Biol. Chem.* 1962, *237* (4), 1375–1376.
40. Arndtsen, B. A.; Bergman, R. G.; Mobley, T. A.; Peterson, T. H., *Acc. Chem. Res.* 1995, *28* (3), 154–162.
41. White, M. C., *Synlett* 2012, *23* (19), 2746–2748.
42. Golisz, S. R.; Gunnoe, T. B.; Goddard, W. A.; Groves, J. T.; Periana, R. A., *Catal. Lett.* 2011, *141* (2), 213–221.
43. Crabtree, R. H., *Chem. Rev.* 1995, *95* (4), 987–1007.
44. Stahl, S. S.; Labinger, J. A.; Bercaw, J. E., *Angew. Chem., Int. Ed.* 1998, *37* (16), 2181–2192.
45. Piera, J.; Backvall, J. E., *Angew. Chem., Int. Ed.* 2008, *47* (19), 3506–3523.
46. Veschetti, E.; Cittadini, B.; Maresca, D.; Citti, G.; Ottaviani, M., *Microchem. J.* 2005, *79* (1–2), 165–170.
47. White, G. C.; Black & Veatch, *White's Handbook of Chlorination and Alternative Disinfectants*, 5th ed.; Wiley: Hoboken, NJ, 2010.
48. Enthaler, S.; Junge, K.; Beller, M., *Angew. Chem., Int. Ed.* 2008, *47* (18), 3317–3321.
49. McDonough, W.; Braungart, M., *Cradle to Cradle: Remaking the Way We Make Things*, 1st ed.; North Point Press: New York, 2002.
50. Breslow, R.; Dong, S. D., *Chem. Rev.* 1998, *98* (5), 1997–2011.
51. Raynal, M.; Ballester, P.; Vidal-Ferran, A.; van Leeuwen, P. W. N. M., *Chem. Soc. Rev.* 2014, *43* (5), 1734–1787.
52. Del Valle, E. M. M., *Process Biochem. (Amsterdam, Neth.)* 2004, *39* (9), 1033–1046.
53. Rideout, D. C.; Breslow, R., *J. Am. Chem. Soc.* 1980, *102* (26), 7816–7817.
54. Chung, W. S.; Wang, J. Y., *J. Chem. Soc., Chem. Commun.* 1995, (9), 971–972.
55. Marchetti, L.; Levine, M., *ACS Catal.* 2011, *1* (9), 1090–1118.
56. Fiedler, D.; Leung, D. H.; Bergman, R. G.; Raymond, K. N., *Acc. Chem. Res.* 2005, *38* (4), 349–358.
57. Leung, D. H.; Bergman, R. G.; Raymond, K. N., *J. Am. Chem. Soc.* 2006, *128* (30), 9781–9797.
58. Leung, D. H.; Bergman, R. G.; Raymond, K. N., *Abstr. Pap. Am. Chem. Soc.* 2004, *227*, U1405.
59. Fiedler, D.; Bergman, R. G.; Raymond, K. N., *Angew. Chem., Int. Ed.* 2004, *43* (48), 6748–6751.
60. Fiedler, D.; van Halbeek, H.; Bergman, R. G.; Raymond, K. N., *J. Am. Chem. Soc.* 2006, *128* (31), 10240–10252.
61. Pluth, M. D.; Bergman, R. G.; Raymond, K. N., *J. Am. Chem. Soc.* 2007, *129* (37), 11459–11467.
62. Pluth, M. D.; Bergman, R. G.; Raymond, K. N., *Science* 2007, *316* (5821), 85–88.
63. Gottesman, S.; Wickner, S.; Maurizi, M. R., *Genes Dev.* 1997, *11* (7), 815–823.
64. Sakamoto, W., *Annu. Rev. Plant Biol.* 2006, *57*, 599–621.
65. Wickner, S.; Maurizi, M. R.; Gottesman, S., *Science* 1999, *286* (5446), 1888–1893.
66. Hill, C. L., *Angew. Chem., Int. Ed.* 2004, *43* (4), 402–404.
67. Hill, C. L.; Delannoy, L.; Duncan, D. C.; Weinstock, I. A.; Renneke, R. F.; Reiner, R. S.; Atalla, R. H. et al. *C. R. Chim.* 2007, *10* (4–5), 305–312.
68. Sloboda-Rozner, D.; Alsters, P. L.; Neumann, R., *J. Am. Chem. Soc.* 2003, *125* (18), 5280–5281.
69. Hill, C. L.; Zhang, X., *Nature* 1995, *373* (6512), 324–326.
70. Nocera, D. G., *Acc. Chem. Res.* 2012, *45* (5), 767–776.
71. Kanan, M. W.; Nocera, D. G., *Science* 2008, *321* (5892), 1072–1075.
72. Lutterman, D. A.; Surendranath, Y.; Nocera, D. G., *J. Am. Chem. Soc.* 2009, *131* (11), 3838–3839.
73. Breslow, R., *J. Biol. Chem.* 2009, *284* (3), 1337–1342.
74. Groves, J. T.; McClusky, G. A., *J. Am. Chem. Soc.* 1976, *98* (3), 859–861.

75. Bell, S. R.; Groves, J. T., *J. Am. Chem. Soc.* 2009, *131* (28), 9640–9641.
76. Groves, J. T.; Lee, J. B.; Marla, S. S., *J. Am. Chem. Soc.* 1997, *119* (27), 6269–6273.
77. Jin, N.; Groves, J. T., *J. Am. Chem. Soc.* 1999, *121* (12), 2923–2924.
78. Jin, N.; Ibrahim, M.; Spiro, T. G.; Groves, J. T., *J. Am. Chem. Soc.* 2007, *129* (41), 12416–12417.
79. Bryliakov, K. P.; Talsi, E. P., *Coord. Chem. Rev.* 2014, *276*, 73–96.
80. Kim, C.; Chen, K.; Kim, J. H.; Que, L., *J. Am. Chem. Soc.* 1997, *119* (25), 5964–5965.
81. Ray, K.; Pfaff, F. F.; Wang, B.; Nam, W., *J. Am. Chem. Soc.* 2014, *136* (40), 13942–13958.
82. Kärkäs, M. D.; Verho, O.; Johnston, E. V.; Åkermark, B., *Chem. Rev.* 2014, *114* (24), 11863–12001.
83. Gloaguen, F.; Rauchfuss, T. B., *Chem. Soc. Rev.* 2009, *38* (1), 100–108.
84. Westerheide, L.; Pascaly, M.; Krebs, B., *Curr. Opin. Chem. Biol.* 2000, *4* (2), 235–241.
85. Koskinen, A., *Asymmetric Synthesis of Natural Products*, 2nd ed.; Wiley: Hoboken, NJ, 2012.
86. De Faveri, G.; Ilyashenko, G.; Watkinson, M., *Chem. Soc. Rev.* 2011, *40* (3), 1722–1760.
87. Jacobsen, E. N.; Palucki, M., Sodium Hypochlorite–N,N′-Bis(3,5-di-tert-butylsalicylidene)-1,2-cyclohexane-diaminomanganese(III) Chloride. In *Encyclopedia of Reagents for Organic Synthesis*. John Wiley & Sons, Ltd: 2001. Available at http://onlinelibrary.wiley.com/doi/10.1002/047084289X.rs085.pub2/abstract.
88. Deng, L.; Jacobsen, E. N., *J. Org. Chem.* 1992, *57* (15), 4320–4323.
89. World Health Organization, 18th WHO Model List of Essential Medicines, 2013. Available at http://www.who.int/medicines/publications/essentialmedicines/18th_EML_Final_web_8Jul13.pdf.
90. 2014 Greener Synthetic Pathways Award. Available at http://www2.epa.gov/green-chemistry/2004-greener-synthetic-pathways-award (accessed November 14, 2014).
91. Kleij, A. W., *Eur. J. Inorg. Chem.* 2009, (2), 193–205.
92. Hanson, J., *J. Chem. Educ.* 2001, *78* (9), 1266.
93. Bauer, A.; Brönstrup, M., *Nat. Prod. Rep.* 2014, *31* (1), 35–60.
94. Nicolaou, K. C.; Yang, Z.; Liu, J. J.; Ueno, H.; Nantermet, P. G.; Guy, R. K.; Claiborne, C. F.; et al. *Nature* 1994, *367* (6464), 630–634.
95. Yamaguchi, J.; Yamaguchi, A. D.; Itami, K., *Angew. Chem., Int. Ed.* 2012, *51* (36), 8960–9009.
96. Chen, M. S.; White, M. C., *Science* 2007, *318* (5851), 783–787.
97. Chen, M. S.; White, M. C., *Science* 2010, *327* (5965), 566–571.
98. Gomez, L.; Canta, M.; Font, D.; Prat, I.; Ribas, X.; Costas, M., *J. Org. Chem.* 2013, *78* (4), 1421–1433.
99. Rosen, B. R.; Simke, L. R.; Thuy-Boun, P. S.; Dixon, D. D.; Yu, J. Q.; Baran, P. S., *Angew. Chem., Int. Ed.* 2013, *52* (28), 7317–7320.
100. Kuhl, N.; Hopkinson, M. N.; Wencel-Delord, J.; Glorius, F., *Angew. Chem., Int. Ed.* 2012, *51* (41), 10236–10254.
101. Haglund, J.; Halldin, M. M.; Brunnstrom, A.; Eklund, G.; Kautiainen, A.; Sandhohn, A.; Iverson, S. L., *Chem. Res. Toxicol.* 2014, *27* (4), 601–610.
102. Zhang, K. Y. E.; Wu, E.; Patick, A. K.; Kerr, B.; Zorbas, M.; Lankford, A.; Kobayashi, T.; Maeda, Y.; Shetty, B.; Webber, S., *Antimicrob. Agents Chemother.* 2001, *45* (4), 1086–1093.
103. McKim, J. M., *Comb. Chem. High Throughput Screen.* 2010, *13* (2), 188–206.
104. Marx, V., *Chem. Eng. News* 2004, *82* (40), 8.
105. Storck, W., *Chem. Eng. News* 2005, *83* (35), 10.
106. U.S. Department of Health and Human Services, Guidance for Industry: Safety Testing of Drug Metabolites, 2008. Available at http://www.fda.gov/OHRMS/DOCKETS/98fr/FDA-2008-D-0065-GDL.pdf.
107. Orhan, H.; Vermeulen, N. P. E., *Curr. Drug Metab.* 2011, *12* (4), 383–394.

108. Cusack, K. P.; Koolman, H. F.; Lange, U. E. W.; Peltier, H. M.; Piel, I.; Vasudevan, A., *Bioorg. Med. Chem. Lett.* 2013, *23* (20), 5471–5483.
109. Andras, B. A.; Kovdari, Z.; Keseru, G. A., *J. Mol. Struc.-Theochem.* 2004, *676* (1–3), 1–5.
110. Balogh, G. T.; Keseru, G. M., *Arkivoc* 2004, *7*, 124–139.
111. Lohmann, W.; Karst, U., *Anal. Bioanal. Chem.* 2008, *391* (1), 79–96.
112. Barret, R.; Pautet, F.; Daudon, M., *Pharmazie* 1987, *42* (2), 132.
113. Bazin, M. J.; Shi, H.; Delaney, J.; Kiine, B.; Zhu, Z. D.; Kuhn, C.; Berlioz, F. et al. *Chem. Biol. Drug Des.* 2007, *70* (4), 354–359.
114. Chorghade, M. S.; Hill, D. R.; Lee, E. C.; Pariza, R. J., *Pure Appl. Chem.* 1996, *68* (3), 753–756.
115. Johansson, T.; Weidolf, L.; Jurva, U., *Rapid Commun. Mass Spectrom.* 2007, *21* (14), 2323–2331.
116. Lipinski, C. A.; Lombardo, F.; Dominy, B. W.; Feeney, P. J., *Adv. Drug Deliv. Rev.* 2001, *46* (1–3), 3–26.
117. Bright, T. V.; Dalton, F.; Elder, V. L.; Murphy, C. D.; O'Connor, N. K.; Sandford, G., *Org. Biomol. Chem.* 2013, *11* (7), 1135–1142.
118. Vulpetti, A.; Dalvit, C., *Drug Discov. Today* 2012, *17* (15–16), 890–897.
119. O'Hagan, D.; Deng, H., *Chem. Rev.* 2015, *115* (2), 634–649.
120. Khetan, S. K.; Collins, T. J., *Chem. Rev.* 2007, *107* (6), 2319–2364.
121. Kirsch, P., *Modern Fluoroorganic Chemistry: Synthesis, Reactivity, Applications*. Wiley-VCH: Weinheim, 2004.
122. Furuya, T.; Kamlet, A. S.; Ritter, T., *Nature* 2011, *473* (7348), 470–477.
123. McMurtrey, K. B.; Racowski, J. M.; Sanford, M. S., *Org. Lett.* 2012, *14* (16), 4094–4097.
124. Liu, W.; Huang, X. Y.; Cheng, M. J.; Nielsen, R. J.; Goddard, W. A.; Groves, J. T., *Science* 2012, *337* (6100), 1322–1325.
125. Kamlet, A. S.; Neumann, C. N.; Lee, E.; Carlin, S. M.; Moseley, C. K.; Stephenson, N.; Hooker, J. M.; Ritter, T., *PLoS One* 2013, *8* (3), e59187.
126. Ellis, W. C.; Tran, C. T.; Roy, R.; Rusten, M.; Fischer, A.; Ryabov, A. D.; Blumberg, B.; Collins, T. J., *J. Am. Chem. Soc.* 2010, *132* (28), 9774–9781.
127. Beach, E. S.; Duran, J. L.; Horwitz, C. P.; Collins, T. J., *Ind. Eng. Chem. Res.* 2009, *48* (15), 7072–7076.
128. Collins, T. J., *Acc. Chem. Res.* 2002, *35* (9), 782–790.
129. Shen, L. Q.; Beach, E. S.; Xiang, Y.; Tshudy, D. J.; Khanina, N.; Horwitz, C. P.; Bier, M. E.; Collins, T. J., *Environ. Sci. Technol.* 2011, *45* (18), 7882–7887.
130. Shappell, N. W.; Vrabel, M. A.; Madsen, P. J.; Harrington, G.; Billey, L. O.; Hakk, H.; Larsen, G. L. et al. *Environ. Sci. Technol.* 2008, *42* (4), 1296–1300.
131. Chanda, A.; Khetan, S. K.; Banerjee, D.; Ghosh, A.; Collins, T. J., *J. Am. Chem. Soc.* 2006, *128* (37), 12058–12059.
132. Chahbane, N.; Popescu, D. L.; Mitchell, D. A.; Chanda, A.; Lenoir, D.; Ryabov, A. D.; Schramm, K. W.; Collins, T. J., *Green Chem.* 2007, *9* (1), 49–57.
133. Kundu, S.; Chanda, A.; Khetan, S. K.; Ryabov, A. D.; Collins, T. J., *Environ. Sci. Technol.* 2013, *47* (10), 5319–5326.
134. Truong, L.; DeNardo, M. A.; Kundu, S.; Collins, T. J.; Tanguay, R. L., *Green Chem.* 2013, *15* (9), 2339–2343.
135. 1999 Academic Award. Available at http://www2.epa.gov/green-chemistry/1999-academic-award (accessed November 14, 2014).

5 Biocatalytic Solutions for Green Chemistry

Erika M. Milczek and Birgit Kosjek

CONTENTS

Biocatalysis is an inherently green methodology providing a valuable alternative for many chemical reactions. Enzymes are proteins that function as biocatalysts and their prominent advantage for green chemistry is that they are renewable, nontoxic, and biodegradable. Enzyme-catalyzed reactions show unprecedented chemoselectivity, regioselectivity, and stereoselectivity often avoiding the need for protection/deprotection steps. The reactions generally do not form side products, which helps to eliminate complex separation and purification procedures. Most biocatalytic processes are run at temperatures slightly above ambient and at near-neutral pH in environmentally benign, mostly aqueous systems. Because biocatalytic processes provide economic and environmental benefits, these processes compete well with chemical methodologies. The best-known attributes and features of biocatalysis clearly coincide with the Twelve Principles of Green Chemistry as outlined by Anastas and Warner (Table 5.1).[1]

Enzymes have been utilized as tools for greener solutions in many industries. The pulp and paper industry has derived advantage from laccase, manganese peroxidase, and other related enzymes for lignin degradation in addition to cellulases to break down cellulosic material. This has significantly improved product quality because cellulose is a naturally occurring contaminant and decreased the environmental impact of the processes.[2] Another example is hydrolytic enzymes commonly used in laundry and dishwashing detergents, stain removers, and industrial/medical cleaning products. These enzymes offer an environmentally friendly solution to degrade protein, carbohydrates, and fats. Moreover, hydrolases with improved performance in cold water allow for additional energy savings by efficiently cleaning clothes at low temperatures. Other examples of widely used biocatalysis applications can be found in the food, agrochemical, petrochemical, fine chemical, and pharmaceutical industries.[2] A common goal of all these areas is to improve efficiency and product output while reducing waste and decreasing its environmental footprint. This is particularly

TABLE 5.1
Biocatalysis Green Chemistry Features

Green Chemistry Principle	Corresponding Attributes of Biocatalysis
1. Waste prevention	Chemoselective enzymes do not form side products avoiding purification steps. Innocuous waste streams often do not require treatment. Biocatalysts are biodegradable
2. Atom economy	Catalytic reaction
3. Less hazardous syntheses	Enzymes not hazardous to human health
4. Designing safer chemicals	Nature created enzymes to be very chemoselective although nontoxic
5. Safer solvents and auxiliaries	Replace organic solvents with aqueous systems
6. Energy efficiency	Near-ambient temperature and pressure
7. Renewable feedstock	Enzymes are produced from glucose
8. Reduce derivatives	Chemospecificity and regiospecificity obviates the use of protecting groups
9. Catalysis	Biocatalysis
10. Design for degradation	Protein and amino acid degradation products pose no problems for waste disposal
11. Real-time analysis	pH change often directly relates to fractional conversion
12. Safer chemistry for accident prevention	Enzymes are safe catalysts, not explosive, flammable, corrosive or toxic

challenging for the pharmaceutical industry due to the complexity of the product molecules and high purity requirements.[3]

This chapter will focus on the most recent enzymatic technologies employed in the fine chemical and pharmaceutical space. Over the past two decades, many diverse enzyme platforms have emerged offering a green alternative to common chemical transformations. Enzymatic asymmetric ketone reductions provide a great example for the efficient production of optically pure alcohols.[4] The use of biocatalysis for hydrolytic reactions can be considered state of the art (examples shown in Schemes 5.1 and 5.2). More recently, the development of transaminases (Scheme 5.7) is another example of a valuable platform that has been applied more and more frequently, revolutionizing the manufacture of optically pure amines.[5] These technologies have significantly affected the design of synthetic routes with an increasing number of enzymatic processes being applied at the manufacturing scale.[6] Other enzyme systems such as laccases, peroxidases, and monoamine oxidases (MAO) are just emerging as practical tools for pharmaceutical processing and offer great opportunities for greener oxidations. The availability of new and useful biocatalysts is

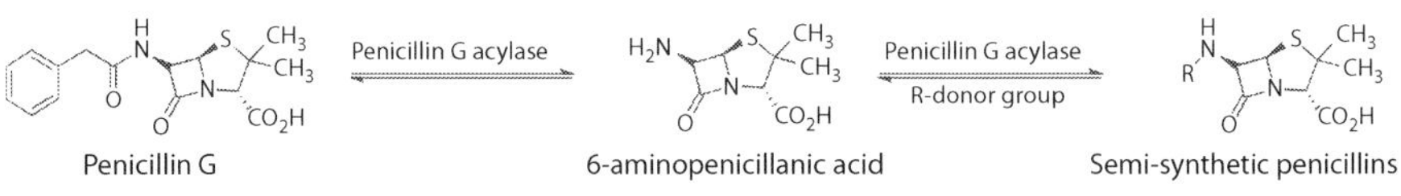

SCHEME 5.1 Penicillin G acylase catalyzed production of semisynthetic β-lactams.

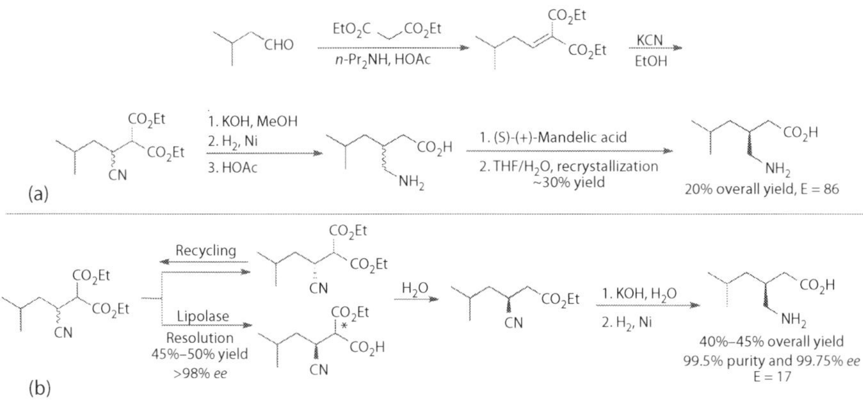

SCHEME 5.2 Pfizer's synthesis of pregabalin. (a) First-generation route featuring a chiral salt crystallization for resolution followed by crystallization of the final amino acid. (b) Biocatalytic route featuring a *T. lanuginosus* lipase catalyzed resolution of a cyano-diester core.

largely a result of advances in enzyme evolution techniques. Recognizing the potential of biotechnology in providing environmentally friendly solutions, the pharmaceutical industry is committed to investing in the further development of enzyme technology.

To further illustrate this point, 20 years ago, most enzymatic reactions were performed either by fermentations of wild-type organisms (growing bulk microorganisms in growth media while converting substrate) or whole cell reactions featuring a recombinant protein. The second option uses resting whole cells produced by fermentation to express the desired enzyme. This required the optimization of fermentation conditions in addition to developing the best parameters for the whole cell transformation. For specific applications, such as when multiple enzymes are needed to regenerate costly cofactors, these systems are still very valuable at the commercial scale. However, the low concentration and throughput of material remains a considerable drawback. Advances in biotechnology have enabled the overexpression of a desired enzyme into host cells such as *Escherichia coli*, making these processes cost-competitive as the host cells provide the required cofactors. However, these reactions are still dilute and can require long optimization timelines that are not conducive for early-stage development in the pharmaceutical industry. In addition, whole cell reactions can pose a significant challenge for isolation and purification of the desired compound.

The transition from whole-cell processes to semipurified isolated enzymes was a major step toward ensuring broad utility in biocatalysis. The two main advantages of using isolated enzymes are (1) clean reactions: use of a single enzyme prevents background reactions that were frequently observed in whole cell transformations; (2) high substrate loading: soluble protein can be engineered using directed evolution (site-directed mutagenesis) to withstand high concentrations of organics allowing for a productive process while minimizing waste. Hydrolases were among the first enzyme reactions to be performed using isolated enzymes rather than whole cells. This is largely due to the availability of diverse wild-type enzymes covering a broad substrate spectrum from enzyme manufacturers. For enzyme systems requiring cofactors, such as ketoreductases, new strategies were developed to carry out *in situ* cofactor recycling without the help of whole cells. Initially, secondary enzyme systems were employed (coupled enzyme approach; Scheme 5.3), but biotechnology soon enabled a more practical coupled substrate approach with a single enzyme performing both the primary reaction and cofactor recycling at the expense of a donor molecule, like low–molecular weight alcohols (Scheme 5.4). In addition, directed evolution technologies have now improved to the extent that enzymes can be evolved to fit desired process parameters including stability, activity, and selectivity. These enzyme improvements are all carried out in the timeline of fully chemical syntheses enabling development timelines competitive with chemical methodologies. This has led to an increasing number of enzymatic processes being included in multistep synthetic routes and has started to reform the conceptual design of total syntheses to active pharmaceutical ingredients (APIs) and fine chemicals toward a more environmentally friendly ideology.

One drawback of these applications is that aqueous waste streams can be contaminated with organics. This waste requires special treatment before disposal, which is costly and can include significant energy consumption. The continuous drive to minimize waste and cut costs has led to an accelerated development of various platforms for enzyme immobilization,[7] of which the effect on manufacturing processes has been recently reviewed.[8] Immobilized enzyme preparations can be used in organic solvents suitable for the downstream chemistry, thus revolutionizing scale-up technologies for biocatalysis by avoiding isolation by extraction and solvent-switching steps. These enzymes can be recovered by simple filtration and, in some cases,

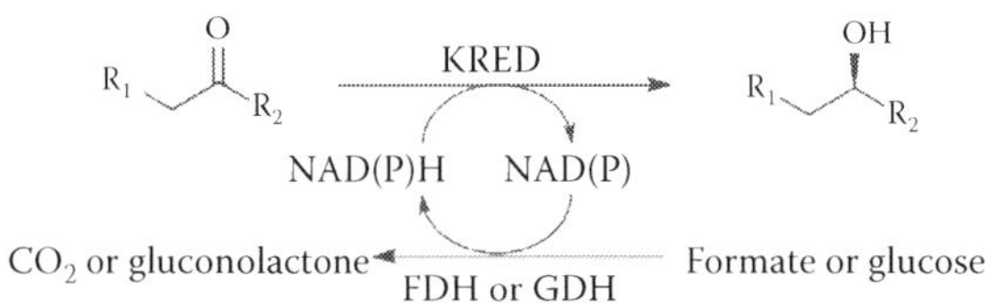

SCHEME 5.3 Classic regeneration systems used for enzymatic ketone reductions featuring GDH or formate dehydrogenase (FDH) and stoichiometric amounts of glucose or formate as a hydride source.

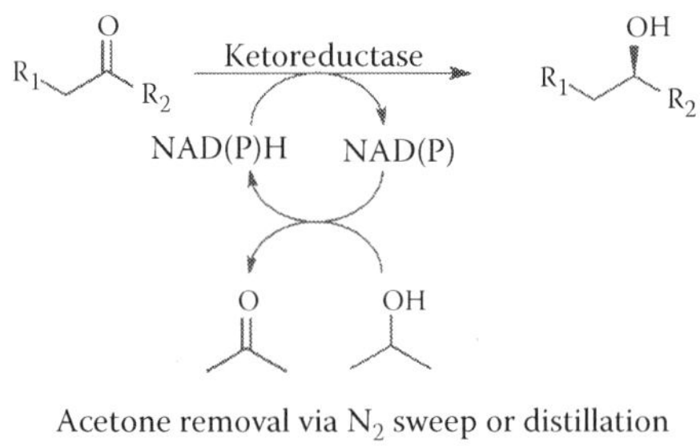

SCHEME 5.4 Industry standard for regenerating NAD(P)H using IPA as a hydride source.

reused in another transformation,[9] enabling greener and more cost-efficient solutions for biocatalytic reactions at the manufacturing scale.

Finally, a measure for improvements in green processing is needed to evaluate different technologies and their effect in increasing efficiency and lowering the environmental footprint. Several mass-based calculations are available for measuring green chemistry performance. For example, E Factor, defined by the ratio of the mass of waste per unit of product, focuses on waste reduction and is one of the most commonly known metrics for green chemistry. Alternatively, process mass intensity (PMI), or the total mass of material required to produce one kilogram of product, highlights efficiency and has recently been favored by the pharmaceutical industry as a leading indicator for green chemistry.[3] In contrast to the E Factor, the PMI sets clear system boundaries and thus is more consistently applicable and comparable across laboratories.

5.1 ESTABLISHED BIOCATALYTIC PLATFORMS

In the pharmaceutical industry, most synthetic endeavors fall under two pillars, either medicinal chemistry or process chemistry. Medicinal chemistry timelines necessitate the rapid generation of synthetic solutions (on the order of days to weeks) requiring an "off the shelf" solution in terms of reaction conditions. Generally speaking, the waste produced in medicinal chemistry syntheses are insignificant because of the small volumes of waste generated at this scale. However, when proof of concept can be established using a biocatalyst in the medicinal chemistry route, a longer timeline is available to optimize and develop the biocatalytic process before large amounts of material are needed to support clinical trials.

In contrast, process chemistry route development benefits from timelines on the order of months to years for viable manufacturing processes. These strategies are therefore amenable to enzyme evolution projects, both internal and external, through contract research organizations (CROs) or academic partners.

Two platforms that have shown the most promise in covering chemical space suitable for both discovery and development endeavors are the hydrolases and ketoreductases. Commercially available enzymes for hydrolysis and acylation cover most of the chemical space for these transformations[10] and both kits and individual enzymes can be obtained. Often, well plates are arrayed with commercially available enzymes and stored until they are required for screening. Limited reaction optimization is required for scaling gram quantities of products. Reaction optimization of these platforms are usually focused on cosolvent, pH, buffer, and other small molecule additives rather than complicated and time-intensive directed evolution strategies. This allows for biocatalytic processes to be implemented in the early discovery space of pharmaceutical research where large numbers of compounds are triaged yearly to find lead candidates for progression through the pipeline.

Hydrolases (nitrilases, esterases, and proteases). The use of hydrolases in asymmetric synthesis is a highly explored field highlighted in a number of reviews featuring deacylation for kinetic resolution and desymmetrization of alcohols and amines,[11,12] asymmetric acylation of alcohols and amines,[13] and asymmetric hydrolysis of nitriles to form carboxylic acids.[14,15] Hydrolases offer a green solution even in kinetic resolution processes as the enzyme often replaces chiral separation such as preparative chiral chromatography, which requires large amounts of solvents. The high enantioselectivity and diastereoselectivity afforded by hydrolases allow for isolated yields to be comparable to separation technologies like chiral supercritical fluid chromatography (SFC). Efficient enzymatic reactions are favored because high substrate loading allows for more efficient solvent use. Penicillin G acylase (PGA) has been one of the most notable examples of biocatalysis affecting the pharmaceutical space (Scheme 5.1).[16]

PGA is used across the industry for the production of 6-aminopenicillanic acid from penicillin G for the synthesis of β-lactam building blocks.[16–20] The extraordinary selectivity of PGA for phenylacetamide-containing functionality has been a benchmark for hydrolytic reactions in the early development of hydrolase platforms as the highly reactive β-lactam functionality remains intact under these mild conditions.

Hydrolases, however, are not limited to the production of β-lactams. Hydrolytic enzymes have been used for biocatalytic transformation by a number of pharmaceutical companies such as GSK,[21–23] Merck,[9,24–26] Pfizer,[27] Novartis,[28] and Roche,[29–31] which illustrates their broad utility in manufacturing processes.

A particularly celebrated example comes from Pfizer in the synthesis of the antiepileptic and neuropathic pain reliever, pregabalin, which is the active ingredient in Lyrica®. The first-generation manufacturing route featured racemic synthesis followed by a classic chiral acid resolution, (S)-mandelic acid (Scheme 5.2a). The acid resolution was later replaced by a biocatalytic route, which utilized Lipolase, a lipase isolated from *Thermomyces lanuginosus*, to resolve the cyano-diester core in 98% *ee* (Scheme 5.2b). Although the chemobiocatalytic route involves a resolution of enantiomers, the E Factor was reduced by an order of five (E Factor = 86 to E Factor = 17) because the use of a biocatalytic resolution allowed for resolution much earlier in the synthesis, and this sequence was amenable to recycling of the undesired isomer.[27] Incorporating the resolution step early in the synthesis allowed for lower waste production in downstream chemistries as the undesired enantiomer was not carried through the subsequent chemical steps. In summary, the improved route reduced energy usage by 82% for total synthesis, further illustrating the favorable environmental impact of the biocatalytic route.

Even greater improvements to biocatalytic processes can be made through immobilization of the enzyme to improve enzyme stability.[32] Packed bed column reactors have been utilized to flush undesirable by-products, which may have inhibitory effects, from the enzyme. This strategy was successfully employed in the asymmetric synthesis of fluoroleucine, a key intermediate in the synthesis of odanacatib.[33] Odanacatib is an inhibitor of cathepsin K, which was developed by Merck for the treatment of osteoporosis. Immobilized *Candida antarctica* lipase B (novozyme-435) was initially used in a stirred tank batch process for the synthesis of the fluoroleucine intermediate; however, using a continuous plugged flow reactor increased catalyst lifetime and minimized side reactions. Moving away from a batch method to a plug flow reactor resulted in an increase from 79% yield and 78% *ee* (batch) to 90% yield and 86% *ee*. The improved outcome of this reaction can be largely attributed to limiting enzymatic inactivation by 20-fold, thereby reducing both cost and waste associated with high enzyme loading in the workup.

Ketoreductases/alcohol dehydrogenases. The ketoreductase (KRED) field is a robust area of research that has proven to be highly fruitful in generating asymmetric solutions in high enantiomeric excess and excellent conversions drawing from enzymes that are in the natural pool. A number of off-the-shelf solutions are commercially available from wild-type organisms and evolved platforms, which is why KREDs are the leading method for asymmetric reduction of ketones in the industry. KREDs have largely replaced chiral boranes and expensive transition metal catalysts due to the plethora of commercially available semipurified enzymes to satisfy most ketone reductions desired by the pharmaceutical industry.[34] The evolution of the KRED field from wild-type whole cell reactions to *E. coli* expression systems followed by isolated semipurified soluble protein has been thoroughly reviewed by Moore and coworkers;[4] therefore, this section will only focus on examples of semipurified isolated KREDs in manufacturing processes. The use of KREDs in the medicinal chemistry (or small molecule discovery) space has increased in popularity in recent years as commercial kits, containing semipurified isolated KREDs, for screening are emerging from enzyme manufacturing companies like Codexis and Johnson Matthey.[35]

This highly fertile field is underscored by numerous examples of enzymatic ketone reductions on the industrial scale. Astra Zeneca,[36] Bristol-Myers Squibb,[37–40] Eli-Lilly,[41,42] Merck,[43,44] and Pfizer[45] have all used KREDs in manufacturing processes to deliver APIs on the kilogram or even ton scale. Unfortunately, the vast majority of these processes are not as "green" as the biotechnology field has come to expect for enzymatic processes. This is largely due to the use of enzyme-coupled recycling systems[46] (such as glucose/glucose dehydrogenase [GDH]; Scheme 5.3) that utilized an additional enzyme as well as a stoichiometric, high–molecular weight second substrate (glucose) as a hydride donor to regenerate NAD(P)H. Although this system demonstrates poor atom economy and requires labor-intensive extraction processes, glucose is a benign and renewable molecule, potentially offsetting such drawbacks. In addition, using the GDH/glucose system allows for real-time analysis through pH shifts in the reaction. This means that the reaction can be stopped when changes in the reaction pH cease.

The use of low–molecular weight alcohols to regenerate NAD(P)H alleviates the poor atom economy associated with enzymatic regenerations systems and glucose as a hydride donor.[47] The most common alcohol regeneration system harnesses isopropanol (IPA) for the regeneration of NAD(P)H by the KRED catalyst employed for reduction of the carbonyl substrate. The equilibrium of this system is further driven by an excess of IPA and removal of the resulting acetone by vacuum or nitrogen sweep (Scheme 5.4).

APIs developed by both Pfizer[48] and BMS[37] have exploited KRED technology for the synthesis of HMG-CoA reductase inhibitors for lowering blood cholesterol. Lipitor® (atorvastatin), marketed by Pfizer, is the first drug to reach annual sales exceeding $10 billion. The motivation to synthesize the API, atorvastatin, is therefore of extreme importance in terms of green and sustainable solutions for such a high volume process. Early in the drug's development, a number of synthetic inefficiencies were identified. An early route featured a sodium borohydride reduction of a β-ketoester in the presence of triethylboron in organic solvent at cryogenic temperatures (Scheme 5.5a). The use of such demanding conditions was reason enough to pursue a greener route; however, 20:1 *cis/trans* selectivity further motivated the development of a more selective and greener approach for the synthesis of the key diol intermediate.

Wild-type ketoreductases were explored and identified, which provided the desired *cis* selectivity.[35] The KRED was expressed in a recombinant form in *E. coli* for ease of overexpression of the protein, and whole cell transformations were explored for the scale-up of the *cis*-diol. Due to the impracticality in shipping frozen whole cells as well as the volume limitation in running whole cell reactions, the desired KRED was isolated as a lysate and stabilized in IPA. The IPA stabilizer served as a cosubstrate for the KRED in the reaction and was used to regenerate the NADH needed for the asymmetric reduction of the ketoalcohol (Scheme 5.5b). This biocatalytic process is more environmentally friendly than the borohydride process as it alleviates the need for cryogenic conditions and reduced organic waste by 65%. Other research groups have further studied the merit of biocatalysts in the production of atorvastatin by exploiting KRED technology in combination with halohydrin dehalogenases (HHDH),[48] and even aldolases[49] (discussed later in this chapter) for the synthesis of the diol core.

Another example that illustrates the utility of KREDs in manufacturing processes is Codexis's synthesis of montelukast, the active ingredient in Singulair.[50] Montelukast is a leukotriene receptor antagonist developed by Merck for the treatment of asthma and seasonal and perennial allergies.[51] Currently, the API is manufactured by Merck using (−)-DIP-Cl for the asymmetric reduction of the key ketone intermediate (Scheme 5.6). Targeting the generic market for montelukast, researchers at Codexis sought to replace this corrosive, moisture-sensitive reductant using an engineered KRED developed through multiple rounds of directed evolution. This novel route boasts high substrate loading (100 g/L) and greater than 97% yield with more than 99.9% *ee* of the chiral alcohol. This route features a much greener process by replacing a chemical process with poor atom economy (borane reductant) and a PMI of 52 with a process using IPA as a hydride donor and PMI of 34. This example demonstrates an exciting new strategy for enzyme engineering companies to compete in the generic pharmaceutical market.

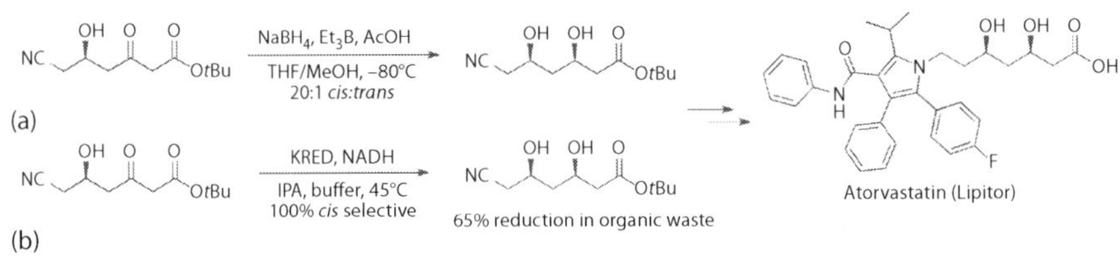

SCHEME 5.5 Comparison of the chemical reduction (a) and biocatalytic route (b) for the production of *cis*-diol intermediate in atorvastatin.

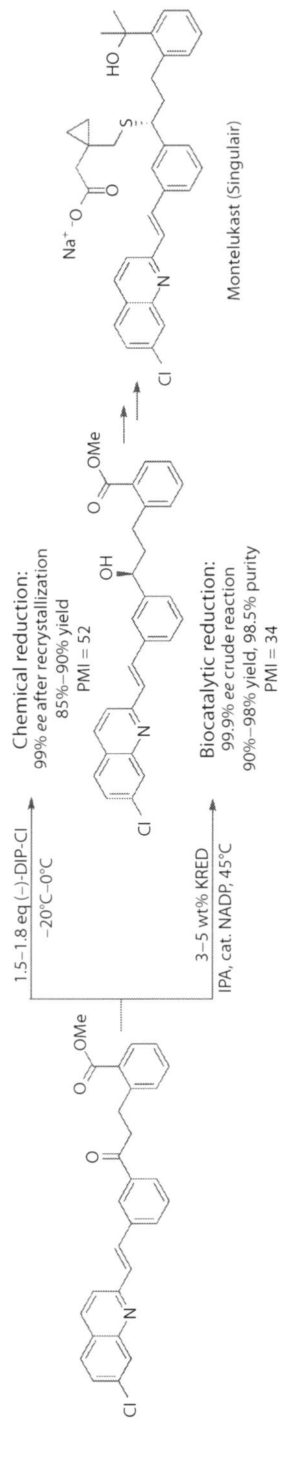

SCHEME 5.6 Codexis strategy for the reduction of key ketone intermediate in the synthesis of montelukast. Comparison of the chemical and biocatalytic routes.

5.2 CUTTING-EDGE BIOCATALYTIC STRATEGIES

Transaminases. Transaminases are a diverse family of enzymes that catalyze the redox neutral group transfer from one molecule to another (Scheme 5.7). Transamination is one of the most provocative new platforms to reach the biocatalysis field due to the chemocatalytic chemistry that transamination seeks to replace, asymmetric reductive amination. Although chemical methods for reductive amination are well developed and readily implemented on the industrial scale, asymmetric versions of this chemistry are plagued with high costs due to the use and removal of precious transition metals and poor safety profiles as a result of the high pressure required for hydrogenations. The use of biocatalytic solutions offers access to inexpensive catalysts, to replace toxic transition metals, and mild reaction conditions as these enzymes perform well at ambient temperature and pressure. Moreover, this technology offers an alternative to the use of hydrogen, a highly flammable gas.

Although the utility and robust nature of ω-transaminases have been realized in manufacturing processes,[52,53] their use in the pharmaceutical setting has been quite limited due to the reversibility of the group transfer reaction as well as the somewhat limited substrate scope (Scheme 5.7). The substrate ketone is replaced by an amine from an amine donor, such as alanine, as shown in Scheme 5.7a. The keto acid resulting from alanine can then react with the transaminase providing the starting ketone and alanine. This reversibility can be partially overcome by using an excess of the amine donor. Alternatively, a cofactor regeneration approach can be explored to remove the keto acid; however, this strategy comes at the expense of adding L-alanine dehydrogenase, NAD, ammonia, and FDH or GDH (Scheme 5.7a). The addition of

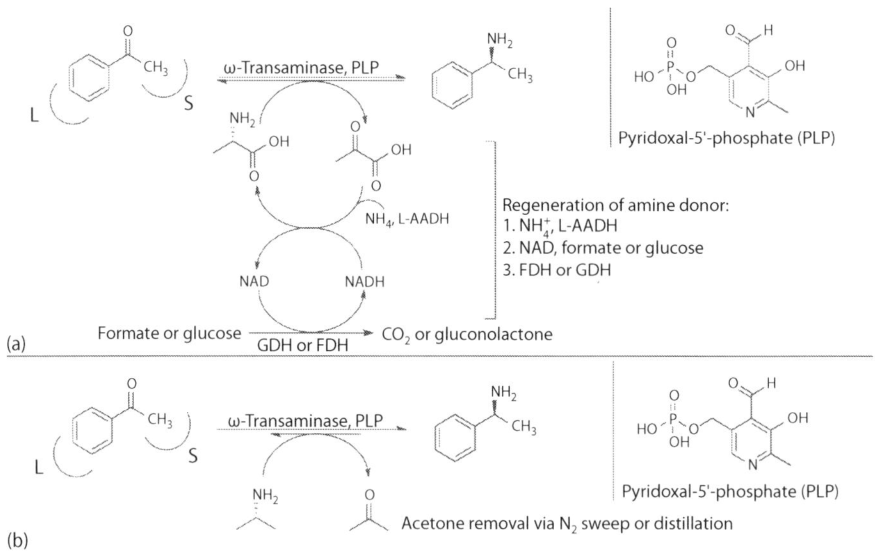

SCHEME 5.7 Standard ω-transaminase reaction: the small (S) and large (L) group allowed in the transaminase active site. (a) Transamination scheme using an amine regeneration system. (b) Transamination scheme using isopropylamine as the amine donor.

stoichiometric glucose/formate makes this methodology less than desirable from a green chemistry and cost of goods standpoint. An alternative method that utilizes isopropylamine (*i*-PrNH_2) is far more attractive as this allows for the removal of the volatile product, acetone, by using a nitrogen sweep to drive the equilibrium in the product direction (Scheme 5.7b). Enzyme engineering has proven to be highly effective in enhancing the activity of transaminases toward isopropylamine as an amine donor.[54,55]

Although the limitations of reversibility have been largely overcome, the substrate scope of these transformations remained a problem until recently.[55] Transaminases generally accommodate a wide variety of simple, low–molecular weight ketone substrates.[56–58] More specifically, this class of enzyme only tolerates groups as large as methyl groups in the "small pocket" (S) of the active site cavity, whereas the large pocket (L) of the active site is more forgiving, accommodating small aliphatic chains and phenyl groups.[59]

Heroic efforts lead by researchers at Codexis and Merck collaboratively solved these limitations by developing a panel of engineered transaminases to accept isopropylamine as the amine donor and a large variety of complicated ketone substrates representative of the substrates seen in the pharmaceutical pipeline.[55] It was demonstrated that the large changes to the architecture of the enzyme active site could not only be tolerated but also that the activity could be competitive with the activities demonstrated on the natural substrates. These challenges were met by synthesizing small model substrates to carve out space from the large active site pocket followed by the small active site pocket over 11 rounds of evolution (Figure 5.1).

This task was framed in the backdrop of the synthesis of sitagliptin (Januvia®), which required transamination of a β-ketoamide core (Scheme 5.8). Sitagliptin is a first in class dipeptidyl peptidase-4 inhibitor (DDP-4) marketed for the treatment of Type II diabetes. Although early chemical routes to sitagliptin were highly efficient and economical with a low E Factor (E = 37), the use of Rh and specialized equipment for high pressure (250 psi hydrogen gas) left room for improvement. Furthermore, the chemical route featured an asymmetric reductive amination that required two crystallization steps to address selectivity issues as well as additional purification to remove residual Rh metals from the final product (Scheme 5.8a). A transamination route could be envisaged from the same β-ketoamide allowing for direct isolation of the sitagliptin-free base eliminating the need for isolation of the enamine intermediate. In addition, a perfectly selective transaminase would also eliminate the need to upgrade the stereochemical purity from 97% *ee*.

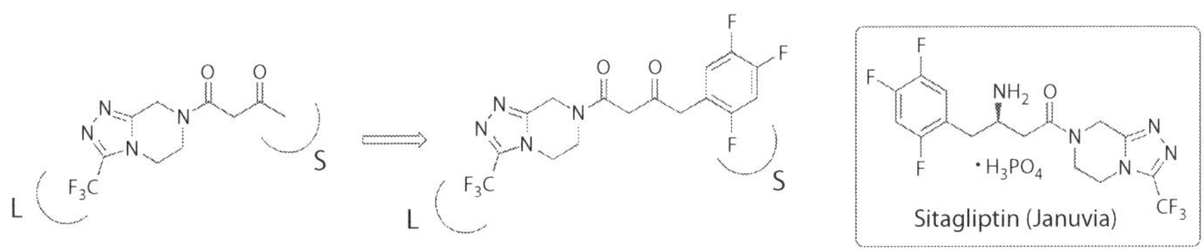

FIGURE 5.1 Designing model substrates for "carving out" space for complex substrates in the transaminase active site.

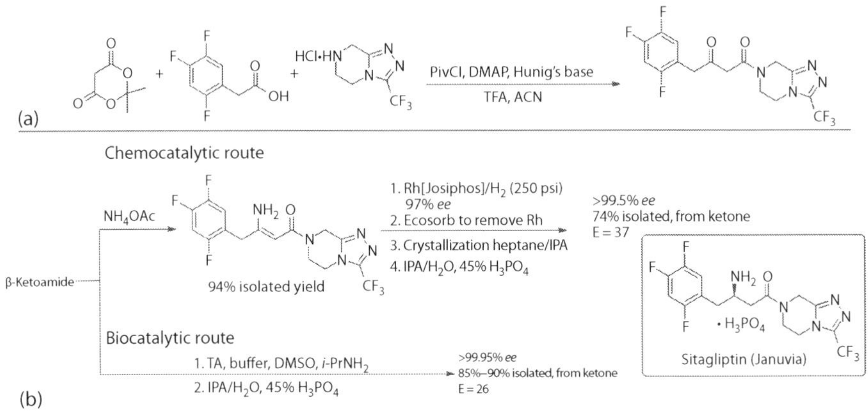

SCHEME 5.8 Synthesis of sitagliptin. (a) Convergent synthesis of the β-ketoamide core. (b) Comparison of the chemocatalytic route and biocatalytic routes to form sitagliptin.

Using the β-ketoamide core, the parent (*R*)-selective-transaminase ATA-117 (a homologue of a ω-transaminase isolated from *Arthrobacter* sp.) showed no activity. Over 11 rounds of evolution, a transaminase was developed to provide the desired enantiomer in greater than 99.95% *ee* and 92% assay yield using 200 g/L β-ketoamide and 6 g/L enzyme loading resulting in an increase in the overall yield by 10% to 13% isolated yield and a 19% reduction in total waste. This process was the recipient of the 2010 Presidential Green Chemistry Award. Note that the chemocatalytic route was awarded the 2006 Presidential Green Chemistry Award because of its highly efficient synthesis with low waste profile (E Factor of 37). The biocatalytic route illustrates that even the best process can be better—featuring an E Factor of 26.

The evolution of this platform has demonstrated such success that many of the compounds that have been evaluated in the Merck pipeline have found a successful solution using a transaminase that can be purchased from the Codexis library (commercially available).[60] This further illustrates the success of the evolution program as well as the overall utility of transaminases in organic synthesis. The next wave of development no doubt resides in the immobilization of this powerful class of enzymes and further stabilization of soluble enzyme to forcing conditions. Academic groups[61–65] as well as the Merck biocatalysis group[66] have demonstrated success in realizing these goals. Although immobilized-TA technology is established, the effect on the field remains to be established. Finally, dynamic kinetic resolution (DKR) using transaminases at high pH leaves yet another avenue for synthetic effects, as this chemistry allows the chemist to set two or more chiral centers in a single transformation.[67,68]

Carbon–carbon bond formation. There are a number of innovative methods for biocatalytic C–C bond formations that have been reviewed recently.[69–71] One of the more explored reactions in pharmaceutical synthesis is the oxynitrilases in which CN^- is added to a carbonyl to prepare optically pure cyanohydrins. The range of these reactions has been recently reviewed.[72,73] Many of these methods have not been demonstrated on the manufacturing scale but have shown promise in the research setting. Examples of such novel enzymatic reactions are the Henry reaction (catalyzed by hydroxynitrile lyases, transglutaminase, and lipases)[74] and Mannich reaction (catalyzed *via* lipases).[75] The remainder of this section will focus on C–C bond–forming reactions that have demonstrated utility on the industrial scale such as amine oxidase and aldolases.

Amine oxidases. Asymmetric amine oxidation for the preparation of imines and iminium ions is an emerging strategy for biocatalytic C–C bond–forming reactions (refer to Scheme 5.10). The flavin-dependent MAO isolated from *Aspergillus niger*, as shown by the Turner laboratory, holds great promise in asymmetric amine oxidation.[76] MAO is responsible for the oxidation of primary and secondary amines to their corresponding imines with the concomitant production of hydrogen peroxide (Scheme 5.9). Catalase can be used to prevent the buildup of hydrogen peroxide in the reaction and subsequent deleterious reactions.

The utility of this methodology was further explored by Schering–Plough (now Merck & Co.) in collaboration with Codexis for the desymmetrization of a key pyrroline derivative in the manufacturing route to boceprevir (Victrelis®).[77] Boceprevir is a first in class NS3 protease inhibitor developed for the treatment of chronic hepatitis C infection by preventing viral replication. A convergent synthesis of boceprevir was envisaged by disconnecting the molecule at the three peptide bonds (Figure 5.2). Synthesis of the dimethylcyclopropylproline methyl ester proved to be challenging. The first-generation route for the synthesis of boceprevir featured a desymmetrization of caronic anhydride to set the two stereocenters followed by eight linear steps, which included a diastereospecific cyanation to set the third chiral center.[78] The second-generation synthesis provided a shorter synthesis by maintaining the symmetry until the final steps.[79–81] In this route, the desired methyl ester diastereomer was isolated by a classic resolution resulting in a 50% loss of material.

Desymmetrization of the pyrrolidine core through asymmetric oxidation of the amine was determined to be the most efficient route; however, dehydrohalogenation

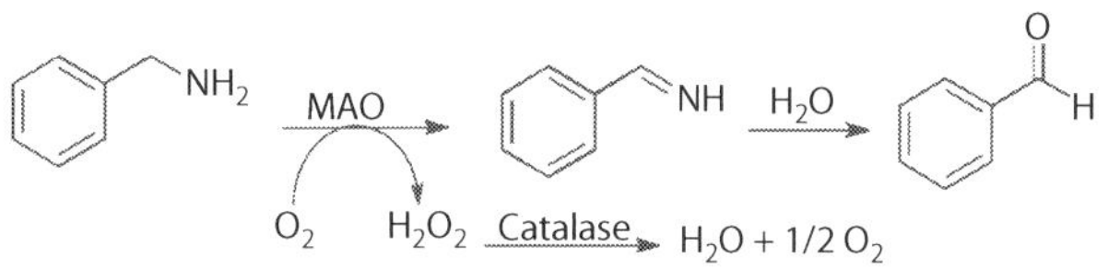

SCHEME 5.9 MAO catalyzed oxidation of benzylamine and disproportionation of hydrogen peroxide catalyzed by catalase.

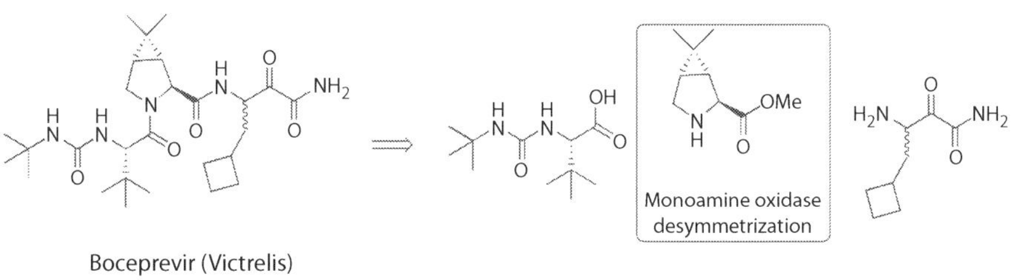

FIGURE 5.2 Retrosynthetic analysis of boceprevir from simple amino acid derivatives.

of the corresponding chloroamine in the presence of phase transfer catalysts afforded only 20% *ee*. After four rounds of directed evolution, MAO provided the desired sulfonate in 95% solution yield and more than 99% *ee* (Scheme 5.10). The bisulfite adduct was directly converted to the corresponding *trans*-nitrile in 90% yield. Under Pinner conditions, the nitrile was then converted to the dimethylcyclopropylproline methyl ester in more than 99% *ee*, more than 99% purity, and 56% isolated yield over the five steps. The biocatalytic route provided a 2.6-fold improvement in yield and eliminated the need for a resolution step. Additionally, the biocatalytic route features molecular oxygen as the terminal oxidant as well as water as the solvent eliminating additional organic waste. Although H_2O_2 is generated in the system, the use of catalase eliminates any safety concerns by converting the peroxide to water and O_2. Furthermore, the imine was found to be a potent inhibitor of MAO on top of being highly volatile, which constitutes a fire hazard when the vapors are mixed with O_2. The current reaction conditions offer a solution by converting imine to the water-soluble sulfonate, thus avoiding the safety hazard. In summary, the biocatalytic route reduced the E Factor by 63% (racemic method E Factor = 191 and biocatalytic method E Factor = 70), illustrating the success of yet another enzymatic platform in pharmaceutical synthesis.

Although examples of MAO catalyzed transformations are limited to this example on a manufacturing scale, the demonstration of this biotransformation as a viable platform for aerobic oxidations is undeniable. The promiscuity of this class of enzymes is still being challenged by academic laboratories,[82,83] and more examples are sure to come as this platform is evolved to accept larger substrates at higher temperatures with greater solvent tolerance.

Aldolases. The use of aldolases in biocatalytic transformations is of particular interest to the biocatalysis community due to the dramatic increase in the molecular weight and complexity of the molecular architecture of the products relative to the simple starting aldehydes and ketones. A particularly provocative example was published by Diversa in which they used deoxyribose-5-phosphate aldolase (DERA) to form the key lactol intermediate used in the synthesis of atorvastatin (Lipitor) and

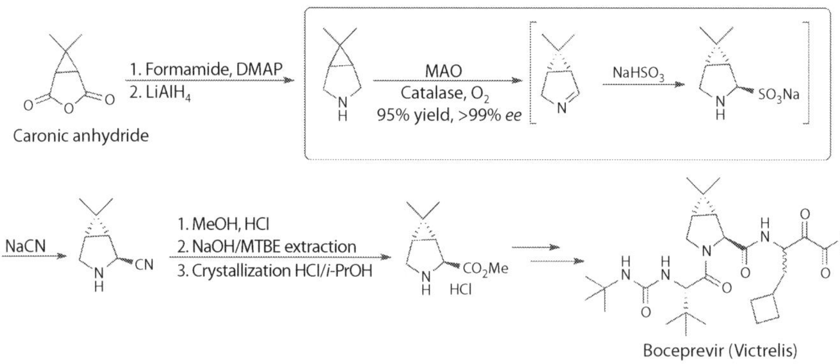

SCHEME 5.10 Biocatalytic route for the production of the dimethylcyclopropylproline methyl ester.

rosuvastatin (Crestor®) (Figure 5.3).[49] The use of DERA in the generic market is of great interest given Lipitor's position as the world's top-selling drug.

Diversa's work was inspired by reports by Wong and colleagues, in which two equivalents of acetaldehyde were coupled using DERA to form four carbon adducts.[84,85] The reversibility of this reaction limits its utility; however, two carbon aldehyde acceptors were used to drive a second condensation forming six-membered lactol products (Scheme 5.11). The stability of the six-membered ring drives the equilibrium and delivers a one-pot tandem aldol reaction that allows two stereocenters to be set in high enantiomeric excess. A variety of chemical functionalities are tolerated on the two aldehydes ranging from halogens to ethers and even azides.

High catalyst loading, long reaction times, and dilute reaction conditions precluded this route from being a viable manufacturing process. Realizing the value of an aldolase-derived process, Diversa screened genomic libraries of environmental DNA obtained from sources around the globe against chloroacetaldehyde and acetaldehyde.[49] High-throughput screening at high substrate concentrations and reduced catalyst load was used to identify a variant of DERA from an unknown organism, which produced the desired lactone in more than 99.9% *ee* and 96.6% *de* after chemical oxidation (Scheme 5.12). Crystallization of the lactone was used to upgrade the diastereomeric excess to 99.8% *de*.

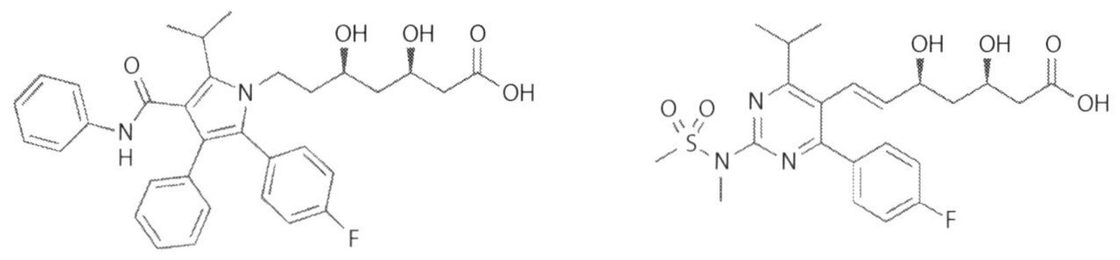

Atorvastatin (Lipitor) Rosuvastatin (Crestor)

FIGURE 5.3 Chemical structures of the cholesterol lowering drugs, atorvastatin and rosuvastatin.

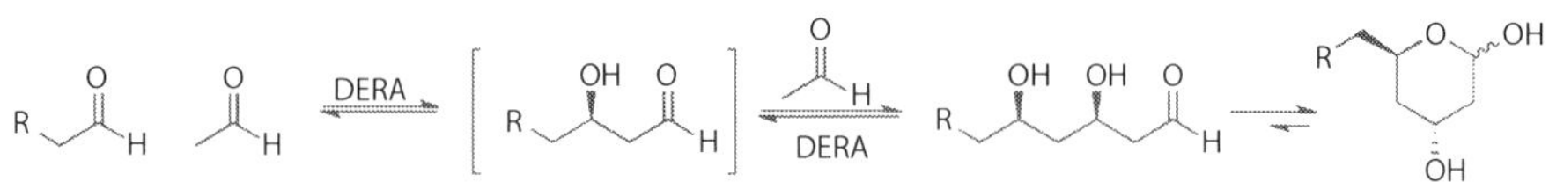

SCHEME 5.11 DERA catalyzed one-pot tandem aldol to form six-membered lactol products.

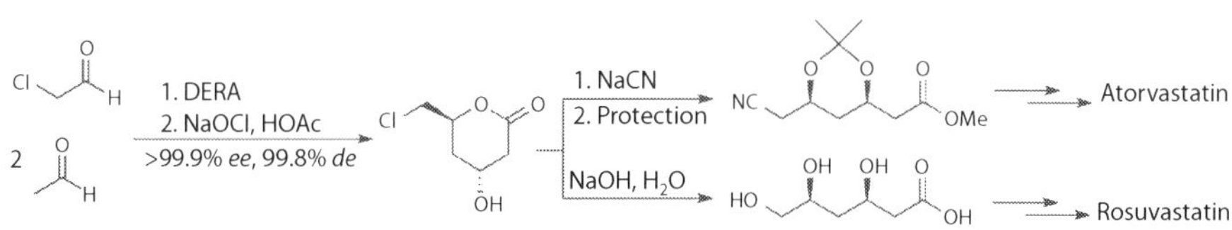

SCHEME 5.12 Use of DERA products to prepare atorvastatin and rosuvastatin.

Because chloroacetaldehyde inhibited the aldolase, a substrate feed was used to keep the chloroacetaldehyde concentrations low. Moving from a batch reaction to a fed-batch process allowed for 380-fold higher volumetric productivity of the reaction (0.08 g/L per hour was improved to 30.6 g/L per hour) and a 10-fold reduction in catalyst loading, which together reduced the overall waste production of this novel process. The aldolase process was performed on a 100 g scale, and the statin side chain was further elaborated to provide the intermediates shown in Scheme 5.12 for the production of atorvastatin (two-step 48% overall yield from the lactol) and rosuvastatin (hydrolysis of the lactone in quantitative yield).

Evaluation of the global genomic pool followed by process optimization allowed for a highly efficient process for the production of the statin side chain in the absence of a directed evolution program. Although aldolases have not had a widespread effect on the pharmaceutical community to date, this technology will no doubt be employed in the years to come.

5.3 EMERGING BIOCATALYTIC STRATEGIES IN THE DEVELOPMENT PHASE

Although the potential of biocatalytic transformations seems endless, a number of challenges remain in moving these platforms toward greener conditions. In the early catalyst development space, the enzymes are often plagued with poor solvent tolerance, temperature instability, low turnover numbers, and intolerance to high concentrations due to the requirement for low substrate concentrations. Many of these limitations can be overcome by enzyme engineering techniques such as directed evolution, enzyme immobilization, and the use of additives to enhance stability or reactivity. However, the starting points for these transformations are often highly wasteful and lack a "green" quality. Already, the biocatalysis field has seen significant developments C–H hydroxylations,[86–91] asymmetric heteroatom oxidations,[92,93] Baeyer-Villiger oxidation,[94–96] enolate reductions,[97] cyanide addition to ketones (oxynitrilases),[98] halogenations (CPO), epoxidations,[99] and group transfer (glycosylation).[100]

5.4 CONCLUSION

A number of processes have been described in this chapter that offer greener solutions than the original fully chemical syntheses. Biocatalytic processes utilizing isolated enzymes often result in high-purity products and alleviate the need for rigorous purification protocols required of transition metal–mediated chemical routes, which can be both costly and wasteful. Even isolated enzyme preparations are met with limitations from a green chemistry perspective. Solvent swapping and laborious extraction from aqueous reactions for subsequent chemistries produce additional waste that must be disposed. The biotechnology field has only just begun to scratch the surface of what is possible by immobilizing enzymes on resins so that these reactions can be run in 100% organic solvent allowing for thorough processing of reaction streams. This would allow the product of the biocatalytic transformation to be isolated by filtration or evaporation of the reaction solvent (a process that is not feasible under aqueous conditions due to low efficiency of evaporating water).

Early biocatalytic processes generally suffered from poor volumetric productivities (high solvent volumes) because high substrate concentrations will often inhibit wild-type enzymes. However, enzyme engineering techniques have largely solved these limitations. Equally exciting is the use of immobilized enzymes for continuous substrate feeds that can also offer work-arounds to these issues when directed evolution is not an option. Enzyme engineering has also offered access to more robust enzymes with greater substrate promiscuity. Through the use of these techniques, the wide-ranging utility of enzymes in pharmaceutical synthesis has been demonstrated. These promising innovations have, in turn, expanded the pool of commercially available biocatalysts. Given the demonstrated benefits of biocatalysts, many researchers continue to engage the broader scientific community through collaboration for the next wave of novel technologies for biocatalysts.

REFERENCES

1. Anastas, P. T., Warner, J. C. *Green Chemistry: Theory and Practice*, Oxford University Press: New York, 1998, 160.
2. Lakshmishri, U., Banerjee, R., Pandey, S. Biocatalysts for greener solutions. *Green Chemistry for Environmental Remediation*. Scrivener Publishing Group LLC.: Beverly, MA, 2012, 479–504.
3. Leahy, D. K., Tucker, J. L., Mergelsberg, I., Dunn, P. J., Kopach, M. E., Purohit, V. C. Seven important elements for an effective green chemistry program: An IQ consortium perspective. *Organic Process Research & Development* 2013, 17 (9), 1099–1109.
4. Moore, J. C., Pollard, D. J., Kosjek, B., Devine, P. N. Advances in the enzymatic reduction of ketones. *Accounts of Chemical Research* 2007, 40 (12), 1412–1419.
5. Savile, C., Mundorff, E., Moore, J. C., Devine, P. N., Janey, J. M. Construction of *Arthrobacter* KNK168 Transaminase Variants for Biocatalytic Manufacture of Sitagliptin. 2010-US25685, 2010099501, 20100226, 2010.
6. Bornscheuer, U. T., Huisman, G. W., Kazlauskas, R. J., Lutz, S., Moore, J. C., Robins, K. Engineering the third wave of biocatalysis. *Nature (London, UK)* 2012, 485 (7397), 185–194.
7. Hanefeld, U., Cao, L., Magner, E. Enzyme immobilisation: Fundamentals and application. *Chemical Society Reviews* 2013, 42 (15), 6211–6212.
8. DiCosimo, R., McAuliffe, J., Poulose, A. J., Bohlmann, G. Industrial use of immobilized enzymes. *Chemical Society Reviews* 2013, 42 (15), 6437–6474.
9. Truppo, M. D., Hughes, G. Development of an improved immobilized CAL-B for the enzymatic resolution of a key intermediate to odanacatib. *Organic Process Research & Development* 2011, 15 (5), 1033–1035.
10. Yazbeck, D. R., Tao, J., Martinez, C. A., Kline, B. J., Hu, S. Automated enzyme screening methods for the preparation of enantiopure pharmaceutical intermediates. *Advanced Synthesis & Catalysis* 2003, 345 (4), 524–532.
11. Reetz, M. T., Jaeger, K. E. Overexpression, immobilization and biotechnological application of *Pseudomonas* lipases. *Chemistry and Physics of Lipids* 1998, 93 (1–2), 3–14.
12. Roberts, S. M., Williamson, N. M. The use of enzymes for the preparation of biologically active natural products and analogs in optically active form. *Current Organic Chemistry* 1997, 1 (1), 1–20.

13. van Rantwijk, F., Sheldon, R. A. Enantioselective acylation of chiral amines catalyzed by serine hydrolases. *Tetrahedron* 2004, 60 (3), 501–519.
14. Tucker, J. L., Xu, L., Yu, W., Scott, R. W., Zhao, L., Ran, N. Modified Nitrile Hydratases and Chemoenzymatic Processes for Preparation of Levetiracetam. 2008-US8503, 2009009117, 20080711, 2009.
15. DeSantis, G., Wong, K., Farwell, B., Chatman, K., Zhu, Z., Tomlinson, G., Huang, H. et al. Creation of a productive, highly enantioselective nitrilase through gene site saturation mutagenesis (GSSM). *Journal of the American Chemical Society* 2003, 125 (38), 11476–11477.
16. Volpato, G., Rodrigues, R. C., Fernandez-Lafuente, R. Use of enzymes in the production of semi-synthetic penicillins and cephalosporins: Drawbacks and perspectives. *Current Medicinal Chemistry* 2010, 17 (32), 3855–3873.
17. Bruggink, A., Roos, E. C., de Vroom, E. Penicillin acylase in the industrial production of β-lactam antibiotics. *Organic Process Research & Development* 1998, 2 (2), 128–133.
18. Matsumoto, K. Production of 6-APA, 7-ACA, and 7-ADCA by immobilized penicillin and cephalosporin amidases. *Bioprocess Technology* 1993, 16 (*Ind. Appl. Immobilized Biocatal.*), 67–88.
19. Rolinson, G. N., Geddes, A. M. The 50th anniversary of the discovery of 6-aminopenicillanic acid (6-APA). *International Journal of Antimicrobial Agents* 2007, 29 (1), 3–8.
20. Sio, C. F., Quax, W. J. Improved β-lactam acylases and their use as industrial biocatalysts. *Current Opinion in Biotechnology* 2004, 15 (4), 349–355.
21. Keeling, S. P., Campbell, I. B., Coe, D. M., Cooper, T. W. J., Hardy, G. W., Jack, T. I., Jones, H. T. et al. Efficient synthesis of an α-trifluoromethyl-α-tosyloxymethyl epoxide enabling stepwise double functionalisation to afford CF3-substituted tertiary alcohols. *Tetrahedron Letters* 2008, 49 (34), 5101–5104.
22. Renganathan, V., Brenner, M. Enzymatic Process for the Production of Cephalosporins. 1997-933517, 6071712, 19970918, 2000.
23. Yu, M. S., Lantos, I., Peng, Z. Q., Yu, J., Cacchio, T. Asymmetric synthesis of (−)-paroxetine using PLE hydrolysis. *Tetrahedron Letters* 2000, 41 (30), 5647–5651.
24. Journet, M., Larsen, R. D., Sarraf, S. T., Shafiee, A., Truppo, M. D., Upadhyay, V. Enzymatic Kinetic Resolution of Racemic Indole Esters. 2004-US14832, 2004104205, 20040512, 2004.
25. Nelson, T. D., LeBlond, C. R., Frantz, D. E., Matty, L., Mitten, J. V., Weaver, D. G., Moore, J. C. et al. Stereoselective synthesis of a potent thrombin inhibitor by a novel P2–P3 lactone ring opening. *Journal of Organic Chemistry* 2004, 69 (11), 3620–3627.
26. Truppo, M. D., Journet, M., Shafiee, A., Moore, J. C. Optimization and scale-up of a lipase-catalyzed enzymatic resolution of an indole ester intermediate for a prostaglandin D2 (DP) receptor antagonist targeting allergic rhinitis. *Organic Process Research & Development* 2006, 10 (3), 592–598.
27. Martinez, C. A., Hu, S., Dumond, Y., Tao, J., Kelleher, P., Tully, L. Development of a chemoenzymatic manufacturing process for pregabalin. *Organic Process Research & Development* 2008, 12 (3), 392–398.
28. Brocklehurst, C. E., Laumen, K., La Vecchia, L., Shaw, D., Vogtle, M. Diastereoisomeric salt formation and enzyme-catalyzed kinetic resolution as complementary methods for the chiral separation of cis-/trans-enantiomers of 3-aminocyclohexanol. *Organic Process Research & Development* 2011, 15 (1), 294–300.
29. Hilpert, H., Wirz, B. Novel versatile approach to an enantiopure 19-nor,des-C,D vitamin D_3 derivative. *Tetrahedron* 2001, 57 (4), 681–694.
30. Wirz, B., Iding, H., Hilpert, H. Multiselective enzymatic reactions for the synthesis of protected homochiral cis- and trans-1,3,5-cyclohexanetriols. *Tetrahedron: Asymmetry* 2000, 11 (20), 4171–4178.

31. Wirz, B., Soukup, M., Weisbrod, T., Staebler, F., Birk, R. Protease-catalyzed preparation of chiral 2-isobutyl succinic acid derivatives for collagenase inhibitor RO0319790. *Asymmetric Catalysis on Industrial Scale*, Part IV. Wiley-VCH Verlag GmbH & Co.: Germany, 2004, 399–411.
32. Adlercreutz, P. Immobilization and application of lipases in organic media. *Chemical Society Reviews* 2013, 42 (15), 6406–6436.
33. Truppo, M. D., Pollard, D. J., Moore, J. C., Devine, P. N. Production of (*S*)-γ-fluoroleucine ethyl ester by enzyme mediated dynamic kinetic resolution: Comparison of batch and fed batch stirred tank processes to a packed bed column reactor. *Chemical Engineering Science* 2007, 63 (1), 122–130.
34. Huisman, G. W., Liang, J., Krebber, A. Practical chiral alcohol manufacture using ketoreductases. *Current Opinion in Chemical Biology* 2010, 14 (2), 122–129.
35. Sutton, P. W., Adams, J. P., Archer, I., Auriol, D., Avi, M., Branneby, C., Collis, A. J. et al. Biocatalysis in the fine chemical and pharmaceutical industries. In: *Practical Methods for Biocatalysis and Biotransformations 2*, John Wiley & Sons, Ltd.: Hoboken, NJ, 2012, 1–59.
36. Holt, R. A., Rigby, S. R. Enzymic Asymmetric Reduction Process to Produce 4H-thieno [2,3-b]thiopyrane Derivatives. 1993-GB1776, 9405802, 19930820, 1994.
37. Patel, R. N., Banerjee, A., McNamee, C. G., Brzozowski, D., Hanson, R. L., Szarka, L. J. Enantioselective microbial reduction of 3,5-dioxo-6-(benzyloxy) hexanoic acid, ethyl ester. *Enzyme and Microbial Technology* 1993, 15 (12), 1014–1021.
38. Patel, R. N., McNamee, C. G., Banerjee, A., Howell, J. M., Robison, R. S., Szarka, L. J. Stereoselective reduction of β-keto esters by *Geotrichum candidum*. *Enzyme and Microbial Technology* 1992, 14 (9), 731–738.
39. Patel, R. N., McNamee, C. G., Banerjee, A., Szarka, L. J. Stereoselective Microbial or Enzymic Reduction of 3,5-Dioxo Esters to 3-Hydroxy-5-Oxo, 3-Oxo-5-Hydroxy, and 3,5-Dihydroxy Esters. 1993-107876, 569998, 19930514, 1993.
40. Patel, R., Hanson, R., Goswami, A., Nanduri, V., Banerjee, A., Donovan, M.-J., Goldberg, S. et al. Enzymatic synthesis of chiral intermediates for pharmaceuticals. *Journal of Industrial Microbiology & Biotechnology* 2003, 30 (5), 252–259.
41. Anderson, B. A., Hansen, M. M., Harkness, A. R., Henry, C. L., Vicenzi, J. T., Zmijewski, M. J. Application of a practical biocatalytic reduction to an enantioselective synthesis of the 5H-2,3-benzodiazepine LY300164. *Journal of the American Chemical Society* 1995, 117 (49), 12358–12359.
42. Vicenzi, J. T., Zmijewski, M. J., Reinhard, M. R., Landen, B. E., Muth, W. L., Marler, P. G. Large-scale stereoselective enzymic ketone reduction with in situ product removal via polymeric adsorbent resins. *Enzyme and Microbial Technology* 1997, 20 (7), 494–499.
43. Chartrain, M., Roberge, C., Chung, J., McNamara, J., Zhao, D., Olewinski, R., Hunt, G., Salmon, P., Roush, D., Yamazaki, S., Wang, T., Grabowski, E., Buckland, B., Greasham, R. Asymmetric bioreduction of 2-(4-nitrophenyl)-*N*-[2-oxo-2-(pyridin-3-yl)ethyl]acetamide to its corresponding (*R*) alcohol [(*R*)-*N*-(2-hydroxy-2-pyridin-3-yl-ethyl)-2-(4-nitro-phenyl)-acetamide] by using *Candida sorbophila* MY 1833. *Enzyme and Microbial Technology* 1999, 25 (6), 489–496.
44. Chartrain, M. M., Chung, J. Y. L., Roberge, C. *N*-(*R*)-(2-Hydroxy-2-Pyridin-3-Ylethyl)-2-(4-Nitrophenyl)Acetamide. 1997-883255, 5846791, 19970626, 1998.
45. Tao, J., McGee, K. Development of a continuous enzymatic process for the preparation of (*R*)-3-(4-fluorophenyl)-2-hydroxypropionic acid. *Organic Process Research & Development* 2002, 6 (4), 520–524.
46. Kosjek, B., Nti-Gyabaah, J., Telari, K., Dunne, L., Moore, J. C. Preparative asymmetric synthesis of 4,4-dimethoxytetrahydro-2H-pyran-3-ol with a ketone reductase and *in situ* cofactor recycling using glucose dehydrogenase. *Organic Process Research & Development* 2008, 12 (4), 584–588.

47. Bradshaw, C. W., Hummel, W., Wong, C. H. *Lactobacillus kefir* alcohol dehydrogenase: A useful catalyst for synthesis. *Journal of Organic Chemistry* 1992, 57 (5), 1532–1536.
48. Ma, S. K., Gruber, J., Davis, C., Newman, L., Gray, D., Wang, A., Grate, J., Huisman, G. W., Sheldon, R. A. A green-by-design biocatalytic process for atorvastatin intermediate. *Green Chemistry* 2010, 12 (1), 81–86.
49. Greenberg, W. A., Varvak, A., Hanson, S. R., Wong, K., Huang, H., Chen, P., Burk, M. J. Development of an efficient, scalable, aldolase-catalyzed process for enantioselective synthesis of statin intermediates. *Proceedings of the National Academy of Sciences of the United States of America* 2004, 101 (16), 5788–5793.
50. Liang, J., Lalonde, J., Borup, B., Mitchell, V., Mundorff, E., Trinh, N., Kochrekar, D. A., Nair Cherat, R., Pai, G. G. Development of a biocatalytic process as an alternative to the (−)-DIP-Cl-mediated asymmetric reduction of a key intermediate of Montelukast. *Organic Process Research & Development* 2010, 14 (1), 193–198.
51. King, A. O., Corley, E. G., Anderson, R. K., Larsen, R. D., Verhoeven, T. R., Reider, P. J., Xiang, Y. B. et al. An efficient synthesis of LTD4 antagonist L-699,392. *Journal of Organic Chemistry* 1993, 58 (14), 3731–3735.
52. Wu, W., Bhatia, M. B., Lewis, C. M., Lang, W., Wang, A., Matcham, G. W. Improvements in the Enzymatic Synthesis of Chiral Amines. 1999-US5150, 9946398, 19990310, 1999.
53. Yamada, Y., Iwasaki, A., Kizaki, N., Ikenaka, Y., Ogura, M., Hasegawa, J. Process for Producing Optically Active Amino Compounds. 1996-JP3054, 9715682, 19961021, 1997.
54. Matcham, G., Bhatia, M., Lang, W., Lewis, C., Nelson, R., Wang, A., Wu, W. Enzyme and reaction engineering in biocatalysis. Synthesis of (*S*)-methoxyisopropylamine (= (*S*)-1-methoxypropan-2-amine). *Chimia* 1999, 53 (12), 584–589.
55. Savile, C. K., Janey, J. M., Mundorff, E. C., Moore, J. C., Tam, S., Jarvis, W. R., Colbeck, J. C., Krebber, A., Fleitz, F. J., Brands, J., Devine, P. N., Huisman, G. W., Hughes, G. J. Biocatalytic asymmetric synthesis of chiral amines from ketones applied to sitagliptin manufacture. *Science (Washington, DC)* 2010, 329 (5989), 305–309.
56. Hoehne, M., Kuehl, S., Robins, K., Bornscheuer, U. T. Efficient asymmetric synthesis of chiral amines by combining transaminase and pyruvate decarboxylase. *ChemBioChem* 2008, 9 (3), 363–365.
57. Truppo, M. D., Rozzell, J. D., Moore, J. C., Turner, N. J. Rapid screening and scale-up of transaminase catalysed reactions. *Organic & Biomolecular Chemistry* 2009, 7 (2), 395–398.
58. Koszelewski, D., Lavandera, I., Clay, D., Guebitz, G. M., Rozzell, D., Kroutil, W. Formal asymmetric biocatalytic reductive amination. *Angewandte Chemie, International Edition* 2008, 47 (48), 9337–9340.
59. Koszelewski, D., Lavandera, I., Clay, D., Rozzell, D., Kroutil, W. Asymmetric synthesis of optically pure pharmacologically relevant amines employing ω-transaminases. *Advanced Synthesis & Catalysis* 2008, 350 (17), 2761–2766.
60. Girardin, M., Ouellet, S. G., Gauvreau, D., Moore, J. C., Hughes, G., Devine, P. N., O'Shea, P. D., Campeau, L.-C. Convergent kilogram-scale synthesis of dual orexin receptor antagonist. *Organic Process Research & Development* 2012, 17 (1), 61–68.
61. Mutti, F. G., Kroutil, W. Asymmetric bio-amination of ketones in organic solvents. *Advanced Synthesis & Catalysis* 2012, 354 (18), 3409–3413.
62. Falus, P., Boros, Z., Hornyanszky, G., Nagy, J., Darvas, F., Uerge, L., Poppe, L. Reductive amination of ketones: Novel one-step transfer hydrogenations in batch and continuous-flow mode. *Tetrahedron Letters* 2011, 52 (12), 1310–1312.
63. Koszelewski, D., Mueller, N., Schrittwieser, J. H., Faber, K., Kroutil, W. Immobilization of ω-transaminases by encapsulation in a sol-gel/celite matrix. *Journal of Molecular Catalysis B: Enzymatic* 2010, 63 (1–2), 39–44.

64. Martin, A. R., Shonnard, D., Pannuri, S., Kamat, S. Characterization of free and immobilized (*S*)-aminotransferase for acetophenone production. *Applied Microbiology and Biotechnology* 2007, 76 (4), 843–851.
65. Yi, S.-S., Lee, C.-W., Kim, J., Kyung, D., Kim, B.-G., Lee, Y.-S. Covalent immobilization of ω-transaminase from *Vibrio fluvialis* JS17 on chitosan beads. *Process Biochemistry (Amsterdam, Netherlands)* 2007, 42 (5), 895–898.
66. Truppo, M. D., Strotman, H., Hughes, G. Development of an Immobilized transaminase capable of operating in organic solvent. *ChemCatChem* 2012, 4 (8), 1071–1074.
67. Chung, C. K., Bulger, P. G., Kosjek, B., Belyk, K. M., Rivera, N., Scott, M. E., Humphrey, G. R., Limanto, J., Bachert, D. C., Emerson, K. M. Process development of C–N cross-coupling and enantioselective biocatalytic reactions for the asymmetric synthesis of niraparib. *Organic Process Research & Development* 2014, 18 (1), 215–227.
68. Peng, Z., Wong, J. W., Hansen, E. C., Puchlopek-Dermenci, A. L. A., Clarke, H. J. Development of a concise, asymmetric synthesis of a smoothened receptor (SMO) inhibitor: Enzymatic transamination of a 4-piperidinone with dynamic kinetic resolution. *Organic Letters* 2014, 16 (3), 860–863.
69. Duenkelmann, P., Mueller, M. Enzymatic C–C coupling in the synthesis of fine chemicals. *Speciality Chemicals Magazine* 2011, 31 (11), 16–18.
70. Fessner, W.-D. Biocatalytic C–C bond formation in asymmetric synthesis. *Asymmetric Synthesis with Chemical and Biological Methods*, Wiley-VCH Verlag GmbH & Co.: Germany, 2007, 351–375.
71. Mueller, M. Recent developments in enzymatic asymmetric C–C bond formation. *Advanced Synthesis & Catalysis* 2012, 354 (17), 3161–3174.
72. Lu, W.-Y., Lin, G.-Q. Chiral synthesis of pharmaceutical intermediates using oxynitrilases. *Biocatalysis for the Pharmaceutical Industry*, John Wiley & Sons (Asia): Singapore, 2009, 89–109.
73. Gotor, V. Biocatalysis applied to the preparation of pharmaceuticals. *Organic Process Research & Development* 2002, 6 (4), 420–426.
74. Milner, S. E., Moody, T. S., Maguire, A. R. Biocatalytic approaches to the Henry (nitroaldol) reaction. *European Journal of Organic Chemistry* 2012, 2012 (16), 3059–3067.
75. Li, K., He, T., Li, C., Feng, X.-W., Wang, N., Yu, X.-Q. Lipase-catalyzed direct Mannich reaction in water: Utilization of biocatalytic promiscuity for C–C bond formation in a "one-pot" synthesis. *Green Chemistry* 2009, 11 (6), 777–779.
76. Alexeeva, M., Enright, A., Dawson, M. J., Mahmoudian, M., Turner, N. J. Deracemization of α-methylbenzylamine using an enzyme obtained by *in vitro* evolution. *Angewandte Chemie, International Edition* 2002, 41 (17), 3177–3180.
77. Li, T., Liang, J., Ambrogelly, A., Brennan, T., Gloor, G., Huisman, G., Lalonde, J. et al. Efficient, chemoenzymatic process for manufacture of the boceprevir bicyclic [3.1.0] proline intermediate based on amine oxidase-catalyzed desymmetrization. *Journal of the American Chemical Society* 2012, 134 (14), 6467–6472.
78. Park, J., Sudhakar, A., Wong, G. S., Chen, M., Weber, J., Yang, X., Kwok, D.-I. et al. Process and Intermediates for the Preparation of (1R,2S,5S)-6,6-Dimethyl-3-Azabicyclo[3,1,0]Hexane-2-Carboxylates or Salts Thereof via Asymmetric Esterification of Caronic Anhydride. 2004-US19135, 2004113295, 20040615, 2004.
79. Wu, G., Chen, F. X., Rashatasakhon, P., Eckert, J. M., Wong, G. S., Lee, H.-C., Erickson, N. C. et al. Process for the Preparation of 6,6-Dimethyl-3-Azabicyclo[3.1.0]Hexane Compounds and Enantiomeric Salts Thereof. 2006-US48613, 2007075790, 20061220, 2007.
80. Berranger, T., Demonchaux, P. Process for Preparation of Optically Pure 6,6-Dimethyl-3-Azabicyclo[3.1.0]Hexane Derivatives. 2007-US25809, 2008082508, 20071218, 2008.

81. Kwok, D.-L., Lee, H.-C., Zavialov, I. A. Dehydrohalogenation Process for Preparation of Intermediates Useful in Providing 6,6-Dimethyl-3-Azabicyclo[3.1.0]Hexane Compounds. 2008-US84174, 2009073380, 20081120, 2009.
82. Ghislieri, D., Houghton, D., Green, A. P., Willies, S. C., Turner, N. J. Monoamine oxidase (MAO-N) catalyzed deracemization of tetrahydro-β-carbolines: Substrate dependent switch in enantioselectivity. *ACS Catalysis* 2013, 3 (12), 2869–2872.
83. Ghislieri, D., Green, A. P., Pontini, M., Willies, S. C., Rowles, I., Frank, A., Grogan, G., Turner, N. J. Engineering an enantioselective amine oxidase for the synthesis of pharmaceutical building blocks and alkaloid natural products. *Journal of the American Chemical Society* 2013, 135 (29), 10863–10869.
84. Gijsen, H. J. M., Wong, C.-H. Unprecedented asymmetric aldol reactions with three aldehyde substrates catalyzed by 2-deoxyribose-5-phosphate aldolase. *Journal of the American Chemical Society* 1994, 116 (18), 8422–8423.
85. Wong, C.-H., Garcia-Junceda, E., Chen, L., Blanco, O., Gijsen, H. J. M., Steensma, D. H. Recombinant 2-deoxyribose-5-phosphate aldolase in organic synthesis: Use of sequential two-substrate and three-substrate aldol reactions. *Journal of the American Chemical Society* 1995, 117 (12), 3333–3339.
86. Conrow, R. E., Harrison, P., Jackson, M., Jones, S., Kronig, C., Lennon, I. C., Simmonds, S., Manufacture of (5Z,8Z,11Z,13E)(15S)-15-Hydroxyeicosa-5,8,11,13-tetraenoic acid sodium salt for clinical trials. *Organic Process Research & Development* 2011, 15 (1), 301–304.
87. Gbewonyo, K., Buckland, B. C., Lilly, M. D. Development of a large-scale continuous substrate feed process for the biotransformation of simvastatin by *Nocardia* sp. *Biotechnology and Bioengineering* 1991, 37 (11), 1101–1107.
88. O'Brien, X. M., Parker, J. A., Lessard, P. A., Sinskey, A. J. Engineering an indene bioconversion process for the production of cis-aminoindanol: A model system for the production of chiral synthons. *Applied Microbiology and Biotechnology* 2002, 59 (4–5), 389–399.
89. Buckland, B. C., Drew, S. W., Connors, N. C., Chartrain, M. M., Lee, C., Salmon, P. M., Gbewonyo, K. et al. Microbial conversion of indene to indandiol: A key intermediate in the synthesis of CRIXIVAN. *Metabolic Engineering* 1999, 1 (1), 63–74.
90. Shibasaki, T., Mori, H., Chiba, S., Ozaki, A. Microbial proline 4-hydroxylase screening and gene cloning. *Applied and Environmental Microbiology* 1999, 65 (9), 4028–4031.
91. Ozaki, A., Mori, H., Shibasaki, T., Ando, K., Chiba, S. Cloning of *Dactylosporangium* Gene for Proline 4-Hydroxylase and Its Use for Fermentative Production of Trans-4-Hydroxy-L-Proline. 1996-709874, 5854040, 19960909, 1998.
92. Baeckvall, J.-E. Selective oxidation of amines and sulfides. In *Modern Oxidation Methods*, 2nd Ed., Wiley-VCH Verlag GmbH & Co.: Germany, 2010, 277–313.
93. Bong, Y. K., Clay, M. D., Collier, S. J., Mijts, B., Vogel, M., Zhang, X., Zhu, J., Nazor, J., Smith, D., Song, S. Engineered Cylohexanone Monooxygenases for Synthesis of Prazole Compounds. 2010-US59398, 2011071982, 20101208, 2011.
94. Doig, S. D., Avenell, P. J., Bird, P. A., Gallati, P., Lander, K. S., Lye, G. J., Wohlgemuth, R., Woodley, J. M. Reactor operation and scale-up of whole cell Baeyer-Villiger catalyzed lactone synthesis. *Biotechnology Progress* 2002, 18 (5), 1039–1046.
95. Alphand, V., Carrea, G., Wohlgemuth, R., Furstoss, R., Woodley, J. M. Towards large-scale synthetic applications of Baeyer-Villiger monooxygenases. *Trends in Biotechnology* 2003, 21 (7), 318–323.
96. Baldwin, C. V. F., Wohlgemuth, R., Woodley, J. M. The first 200-L scale asymmetric Baeyer-Villiger oxidation using a whole-cell biocatalyst. *Organic Process Research & Development* 2008, 12 (4), 660–665.

97. Bechtold, M., Brenna, E., Femmer, C., Gatti, F. G., Panke, S., Parmeggiani, F., Sacchetti, A. Biotechnological development of a practical synthesis of ethyl (*S*)-2-ethoxy-3-(*p*-methoxyphenyl)propanoate (EEHP): Over 100-fold productivity increase from yeast whole cells to recombinant isolated enzymes. *Organic Process Research & Development* 2011, 16 (2), 269–276.
98. Roberge, C., Fleitz, F., Pollard, D., Devine, P. Synthesis of optically active cyanohydrins from aromatic ketones: Evidence of an increased substrate range and inverted stereoselectivity for the hydroxynitrile lyase from *Linum usitatissimum*. *Tetrahedron: Asymmetry* 2007, 18 (2), 208–214.
99. Panke, S., Held, M., Wubbolts, M. G., Witholt, B., Schmid, A. Pilot-scale production of (*S*)-styrene oxide from styrene by recombinant *Escherichia coli* synthesizing styrene monooxygenase. *Biotechnology and Bioengineering* 2002, 80 (1), 33–41.
100. Auriol, D., ter Halle, R., Lefèvre, F., Visser, D. F., Gordon, G. E. R., Bode, M. L., Mathiba, K. et al. Transferases for alkylation, glycosylation and phosphorylation. In: *Practical Methods for Biocatalysis and Biotransformations 2*, John Wiley & Sons, Ltd.: Hoboken, NJ, 2012, 231–262.

6 Montmorillonite Clays as Heterogeneous Catalysts for Organic Reactions

Matthew R. Dintzner

As concerns for society, the environment, and the economy continue to shape the way chemists think about the construction of small molecules, the development of synthetic methodologies that promote sustainability is essential (for a recent review, see Ref. 1).[2] Environmentally benign, naturally abundant clays are ideally suited not only for the "greening" of modern synthesis but also for incorporating into the repertoires of a new generation of synthetic chemists for whom sustainability is a growing priority.[3] The advantages of using Montmorillonite clays are many:

- Clays are an abundant and benign natural resource
- Clays are inexpensive and commercially available
- Clays are easy to use and safe to handle
- Clays are exceedingly versatile in that their activity may be attenuated through heating, pillaring, or cation exchanging

Although Mother Nature's synthetic prowess has yet to be matched by man, the challenges of unraveling the mysteries of nature have led to the development of a most impressive body of knowledge. Some have speculated that the molecules of life originated through clay-mediated reactions.[4] Whether or not this is the case, the elegance of harnessing nature's resources to manipulate matter at the molecular level is undeniable. Additionally, in emulating nature, we are ever more diligent in our efforts to preserve the environment that sustains us. Toward that end, we have endeavored to develop a research program that readily engages undergraduate students in the scientific method while also expanding the scope of reactions that may be successfully affected under milder conditions—namely, by replacing traditional, caustic Bronsted and Lewis acid catalysts with Montmorillonite clays.

Fundamental to the art and science of synthetic organic chemistry is the ability to control the formation of new bonds, especially new carbon–carbon bonds. Although it is challenging to alter the thermodynamics of a reaction, controlling the kinetics, or rate, of a reaction is straightforward. One method that chemists have relied on to increase the rate of a chemical reaction is through the use of a catalyst. A catalyst is a species that increases the rate of a reaction but is itself not consumed in the reaction. Thus, catalysts contribute to sustainability insofar as they may be recycled and reused.

This is especially true of naturally abundant clay minerals, like Montmorillonites, which have been shown to effectively catalyze a broad array of organic reactions, including a number of important carbon–carbon bond–forming reactions.[1,2]

In the most general sense, clays are a type of fine-grained earth, primarily composed of extensively layered aluminum and silicate minerals.[5] Montmorillonite clays are thought to have formed from volcanic ash during the Jurassic and later periods and were named for the location of their discovery, Montmorillon, France, in the 1800s. Montmorillonite clays are now mined from regions all over the world, including Europe, Africa, Asia, and South and North America. In addition, acid-treated clays, such as Montmorillonite K10 and Montmorillonite KSF, are available through manufacturers like Sigma-Aldrich in the United States, and have been used extensively in organic synthesis.[1,2,6]

Montmorillonite K10, which is a member of the smectite family of minerals, has a layered structure that consists of sheets of tetrahedral silicate $\left(SiO_4^{4-}\right)$ ions that sandwich sheets of octahedral aluminum (Al^{3+}) oxide ("T-O-T" scaffold; Figure 6.1), forming an interconnected network via weak bonds with oxygen. Natural, intermittent substitutions of Al^{3+} or Fe^{3+} for Si^{4+}, or Fe^{2+} or Mg^{2+} for Al^{3+} (M; Figure 6.1), result in a charge imbalance that is compensated for by intercalation of alkaline earth metal cations like Na^+ and Mg^{2+}, and water, in the spaces between the layers (interlayer region; Figure 6.1).[5] These structural features impart some Lewis and Bronsted acidic properties to the clay that may be exploited in catalysis.[6]

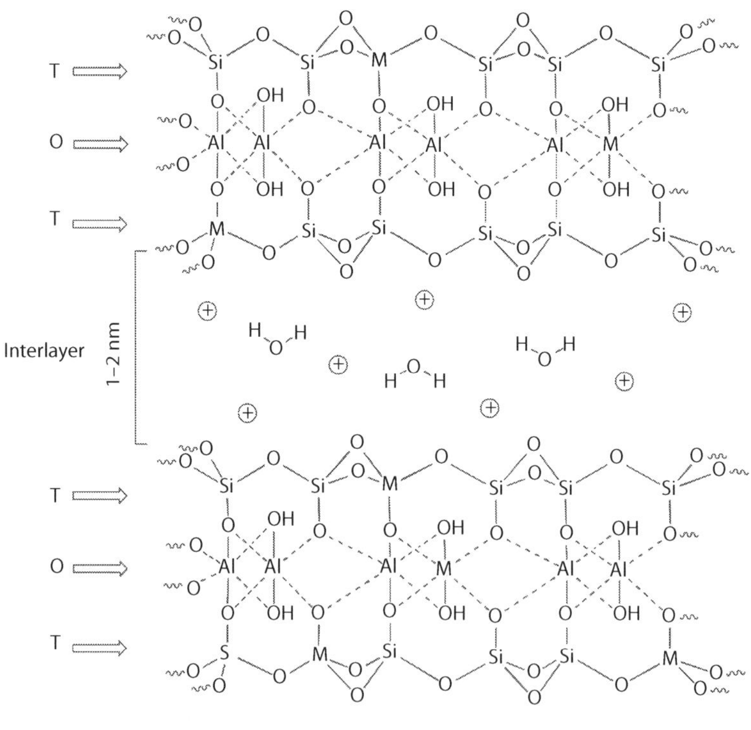

FIGURE 6.1 Layered structure of Montmorillonite clay.

One useful property of clays with regard to their application in promoting sustainability is that they are insoluble solids and, as such, function as heterogeneous catalysts (for a review, see Ref. 7). Heterogeneous catalysts differ in phase from the reactants and reaction medium; often, the catalyst is a solid and the reactants are in the gas or solution phase. Although the catalyst and reactant molecules comprise different phases, they nonetheless interact noncovalently, the catalyst providing a surface upon which the reactants are adsorbed. Adsorption of the reactants onto the surface of the catalyst facilitates the interaction of reactants with one another, thus promoting the reaction. Product molecules then desorb from the surface of the catalyst, freeing up catalytic sites for subsequent reactions. The surface area of the catalyst, therefore, is an important consideration in heterogeneous catalysis because increased surface area results in greater catalytic activity.[7]

As heterogeneous solid catalysts, clays are readily separated by filtration from liquid product mixtures at the end of a reaction, obviating the need for elaborate workup and purification procedures that would otherwise generate considerable waste. This also makes it convenient to recover the clay material, which may often be reused many times in succession with only minimal diminishment in catalytic activity. On the other hand, the clay's heterogeneity makes it somewhat more difficult to elucidate reaction mechanisms and optimize conditions. Furthermore, clay-catalyzed reactions tend to be much more capricious than homogeneous reactions with regard to substrate tolerance, which can significantly narrow the scope of their applications. It is precisely the challenges of overcoming these obstacles, and the rewards of contributing to sustainability, that fuels our efforts to investigate clays as catalysts for organic reactions and expand on their applications in green chemistry.

In the last decade, several excellent reviews have highlighted burgeoning developments in the application of clays as catalysts for organic reactions.[1,2,6] Although these studies will not be repeated here, an account of our own experiences with incorporating clays into an undergraduate research program for which sustainability is a primary concern will be presented. Many of the goals of sustainability, in fact, are well aligned with developing a successful and productive undergraduate research laboratory program. Reducing cost and working with materials that are safe to handle, for example, are not only integral to the theme of sustainability but also allow for meaningful research to be done by novices on a budget.

We first endeavored to use clay in the context of a project aimed at the synthesis of an anti-inflammatory acetophenone glucoside natural product (**1**).[8] The key step in our synthetic plan featured a clay-catalyzed, sigmatropic [1,3]-shift reaction (**3** → **2**; Scheme 6.1).

Although preparation of **3** from commercially available 4′-hydroxyacetophenone was straightforward, treatment of **3** with Montmorillonite KSF clay, according to the method of Dauben and coworkers,[9] resulted in no reaction. At the time, this was not the desired result but it prompted us to investigate Dauben's protocol in greater depth. According to Dauben, treatment of allyl phenyl ether **4** in benzene with one equivalent (by mass) of Montmorillonite KSF clay over the course of 9 h gave *o*-prenyl phenol **5a** in 34% yield, along with trace amounts of the *para* isomer **5b** and heterocycle **5c** (Scheme 6.2).

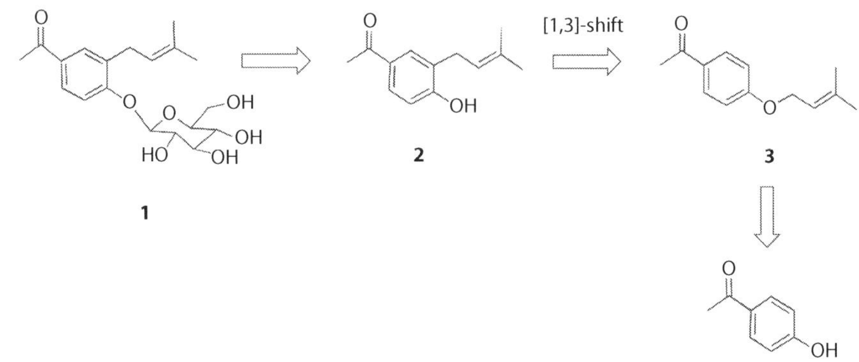

SCHEME 6.1 Synthetic plan featuring a clay-catalyzed [1,3]-shift reaction.

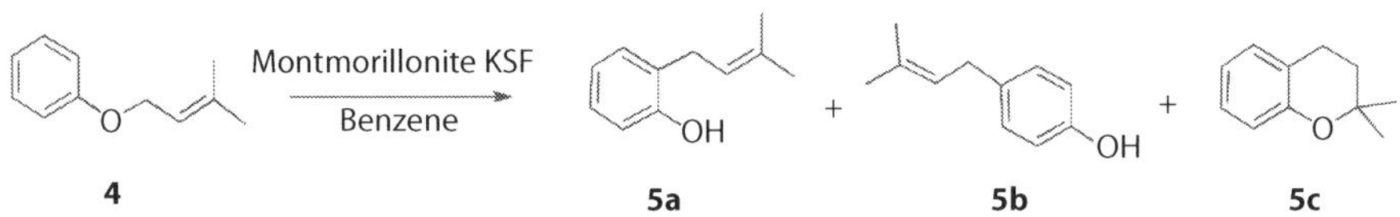

SCHEME 6.2 Clay-catalyzed [1,3]-shift reaction.

We repeated this experiment and observed similar results. However, when we used Montmorillonite K10 clay (purchased from Sigma-Aldrich for pennies per gram!) in place of the KSF, we observed a significant increase in the rate of reaction, presumably due to the increase in surface area of K10 relative to KSF. Intrigued by this result, we probed other experimental variables (solvent, temperature, and time) and were encouraged to find that we were able to achieve a reasonable degree of control over the reaction and optimize its conditions.[10] In the end, we doubled the yield of **5a** simply by using a more active catalyst (Montmorillonite K10) in a different solvent (CCl_4). Although these conditions were still ineffective for our originally planned synthetic scheme (**3** → **2**), we were nonetheless intrigued by the activity of the clay and set out on a course to uncover more ways that we might exploit its properties.

One interesting observation we made during the previous investigation was that prolonged reaction times resulted in the formation of greater amounts of the cyclized product (**5c**). As a multitude of natural products, including coumarins, chromenes, and flavonoids, contain this 2,2-dimethylbenzopyran moiety as part of their carbon framework, we elected to further investigate **5c** as a synthetic target.[11] We found that these compounds could be directly generated from the corresponding phenols (**6c**) and prenyl bromide in the presence of Montmorillonite K10 clay in CCl_4 at ambient temperature (Scheme 6.3).[12]

Although we were reasonably satisfied with our results (Table 6.1), having developed a more environmentally friendly and sustainable route to the target compounds, we were vexed by the variability in percent conversions as a function of the substrate. Why was it that the reaction proceeded quantitatively with 2,3-dimethoxyphenol

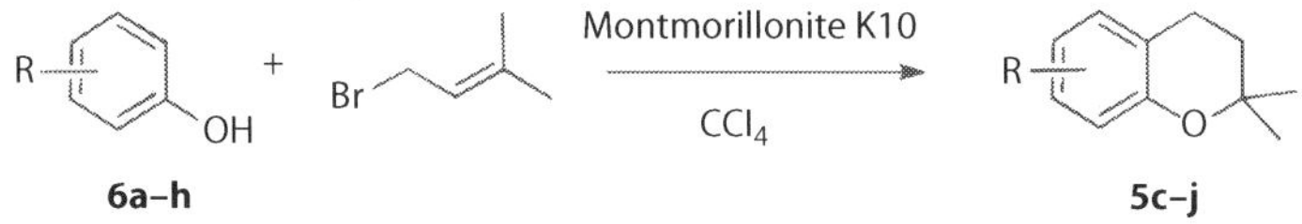

SCHEME 6.3 Clay-catalyzed synthesis of 2,2-dimethylbenzopyrans.

TABLE 6.1
Scope of the Clay-Catalyzed Reaction of Phenols with Prenyl Bromide

Entry	Phenol (6)	Product	% Conversion
1	Phenol (**6a**)	**5c**	67
2	3,4-Methylenedioxyphenol (**6b**)	**5d**	91
3	3,4-Dimethoxyphenol (**6c**)	**5e**	100
4	4-Methoxyphenol (**6d**)	**5f**	84
5	4-Methylphenol (**6e**)	**5g**	91
6	4′-Hydroxyacetophenone (**6f**)	**5h**	76
7	4-Hydroxybenzaldehyde (**6g**)	**5i**	41
8	4-Nitrophenol (**6h**)	**5j**	0

(Table 6.1, entry 3), and not at all with *p*-nitrophenol (Table 6.1, entry 8)? These data may have suggested a simple substituent effect, but percent conversions were comparable for *p*-methoxyphenol and 4′-hydroxyacetophenone (Table 6.1, entries 4 and 6, respectively), which did not follow.[12]

Expanding on this work, we carried out a two-step synthesis of methylenedioxyprecocene (MDP, **7**; Scheme 6.4), a natural product that exhibits anti-juvenile hormone activity in some insects.[13]

Although the first step in the above synthesis was both high yielding and environmentally friendly, the second was anything but.[14] In an effort to circumvent this shortcoming, we developed a one-step synthesis of MDP that involved the neat, microwave assisted, Montmorillonite clay-catalyzed condensation of sesamol (**6b**) with 3-methyl-2-butenal to give **7** directly (Scheme 6.5).[15]

The above scheme is a vast improvement over our original, not only because it reduces the number of steps from two to one but also because it uses a more efficient energy source (microwave irradiation), involves no halogenated organics (reagents or solvents), and generates only water as a by-product. This reaction was, in fact, so

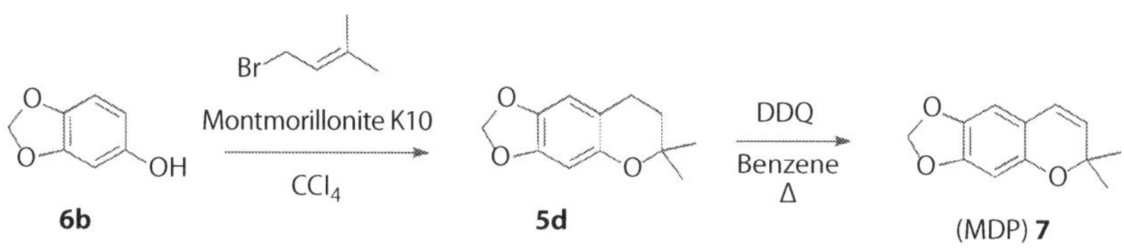

SCHEME 6.4 Two-step synthesis of MDP.

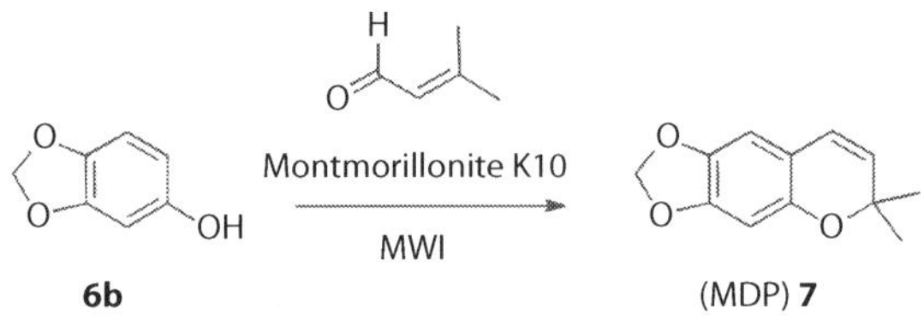

SCHEME 6.5 One-step synthesis of MDP.

clean and simple to execute that it was adapted as a laboratory experiment for the undergraduate organic chemistry curriculum.[16]

By way of comparison, consider an alternative synthesis of chromene-type compounds, reported in the literature over 20 years ago: reaction of α,β-unsaturated aldehydes with phenols in the presence of phenylboronic acid (Scheme 6.6).[17]

Although this procedure was observed to generate the corresponding chromenes in good yields, the methodology falls short of ours on several counts: (1) phenylboronic acid is considerably more expensive than Montmorillonite clay (by orders of magnitude!), (2) the reaction is conducted in benzene, (3) the reaction requires conventional heating, (4) the reaction requires aqueous workup, (5) phenyl borate is generated as a waste product.

At this point in the development of our research program, we were beginning to acquire a good sense of the strengths and limitations of the clay, and we next turned our attention to investigating the clay-catalyzed hetero-Diels–Alder reaction of 2,3-dimethyl-1,3-butadiene (**8**) with aldehyde dienophiles (**9**) to give heterocyclic product **10** (Scheme 6.7).[18] The Diels–Alder reaction is arguably one of the most elegant, extensively studied, and widely used reactions in synthetic organic chemistry. Additionally, it is by nature a "green" reaction in that it can potentially proceed with perfect atom economy—every atom of the starting materials may be incorporated into the product(s). Although Diels–Alder reactions are often spontaneous, hetero-Diels–Alder reactions typically require a Lewis acid catalyst (M; Scheme 6.7) to compensate for the dienophile's diminished electrophilicity.

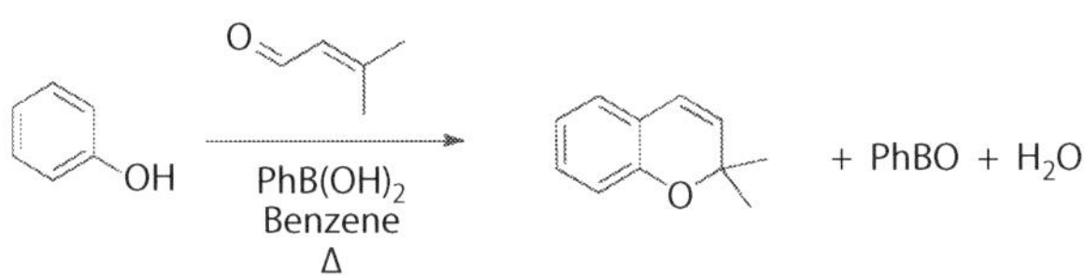

SCHEME 6.6 Alternative chromene synthesis.

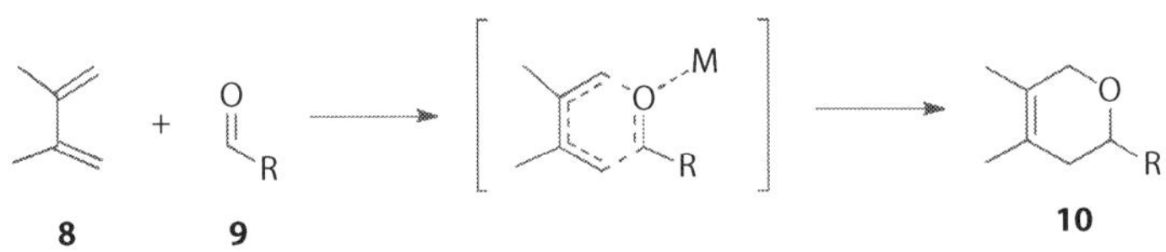

SCHEME 6.7 Hetero-Diels–Alder reaction.

A whole host of Lewis acid catalysts have been developed for successfully affecting the hetero-Diels–Alder reaction, including many that allow for some degree of stereoselectivity (for a review, see Ref. 19). We therefore felt that the prospect of successfully catalyzing these sorts of reactions with Montmorillonite clay was promising. Given our observations from previous studies, we anticipated that electron-deficient dienophiles like nitrobenzaldehydes would perform best, whereas electron-rich substrates such as anisaldehydes would perform most poorly. The results of our investigation, however, were surprising (Table 6.2), especially with regard to the highest-yielding reaction—that of *o*-anisaldehyde (**9b**).[18]

Given the strongly electron-donating nature of the methoxy substituent, it was expected that the reaction of **9b** would proceed in very low yield (if at all), as was the case with *p*-anisaldehyde (**9j**). The fact that the reaction proceeded best with **9b** was curious, indeed, and we rationalized that it was the result of a unique interaction between this, and similar substrates, and the clay, presumably lowering the activation energy of the reaction's transition state (**11**; Figure 6.2).

Although interesting from a mechanistic perspective, we found the clay-catalyzed hetero-Diels–Alder reaction to be less useful from the perspective of

TABLE 6.2
Scope of the Clay-Catalyzed Hetero-Diels–Alder Reaction

Entry	Aldehyde (9)	Product	% Conversion
1	Benzaldehyde (**9a**)	**10a**	52
2	*o*-Anisaldehyde (**9b**)	**10b**	81
3	*o*-Ethoxybenzaldehyde (**9c**)	**10c**	25
4	*o*-Fluorobenzaldehyde (**9d**)	**10d**	49
5	*o*-Chlorobenzaldehyde (**9e**)	**10e**	34
6	*o*-Bromobenzaldehyde (**9f**)	**10f**	68
7	*o*-Cyanobenzaldehyde (**9g**)	**10g**	22
8	*o*-Nitrobenzaldehyde (**9h**)	**10h**	75
9	*o*-Tolualdehyde (**9i**)	**10i**	7
10	*p*-Anisaldehyde (**9j**)	**10j**	<1
11	*p*-Chlorobenzaldehyde (**9k**)	**10k**	56
12	*p*-Nitrobenzaldehyde (**9l**)	**10l**	56

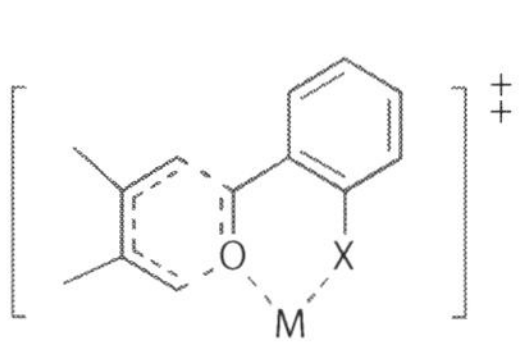

11

X = $-OCH_3$, $-F$, $-Cl$, $-Br$, $-OCH_2CH_3$

FIGURE 6.2 Proposed hetero-Diels–Alder transition state.

developing a more sustainable chemical process. The scope of the reaction seemed to be very narrow and substrate-specific, not only with regard to the aldehyde component but also with respect to the diene. Attempts to carry out the reaction with less substituted dienes, such as isoprene, were only minimally successful. Furthermore, when we attempted to expand the scope of the reaction by using aliphatic aldehydes (**12**), we observed minimal formation of Diels–Alder adducts due to a competing side reaction: cyclotrimerization to give 1,3,5-trioxanes (**13**; Scheme 6.8).

Because 1,3,5-trioxanes have been implicated in a host of practical applications, and because their synthesis in the presence of Montmorillonite clay had not been reported, we were encouraged to further investigate this reaction and attempt to optimize its conditions. Unfortunately, what we found was that this reversible reaction was exceedingly difficult to control, despite extensive efforts to do so.[20] In the course of our investigation, however, we observed another competing reaction: treatment of aliphatic aldehydes with Montmorillonite KSF clay, neat, resulted in irreversible oxidation to the corresponding carboxylic acids (**14**; Scheme 6.9).[20]

We found that this reaction proceeded slowly at ambient temperature in good to excellent yields (Table 6.3), utilizing atmospheric oxygen as the oxidant. This procedure constitutes a much more sustainable alternative to traditional methodologies for the synthesis of aliphatic carboxylic acids.[21] The relative toxicity of some of the more commonly used reagents for oxidation of aldehydes (peroxides, $KMnO_4$, oxone, etc.) far exceeds that of our clay/air reaction. Furthermore, our system is much easier and more convenient to execute, considerably more economical, and generates far less waste.

To date, perhaps the most successful reaction to have been developed in our laboratories, at least with regard to sustainability, is the Montmorillonite K10 clay–catalyzed Hosomi–Sakurai-type reaction.[22] Generally considered among the most useful carbon–carbon bond-forming reactions in natural products synthesis, we were delighted to observe that treatment of aromatic aldehydes (**9**) with allyltrimethyl silane in the presence of Montmorillonite K10 clay gave near-quantitative yields

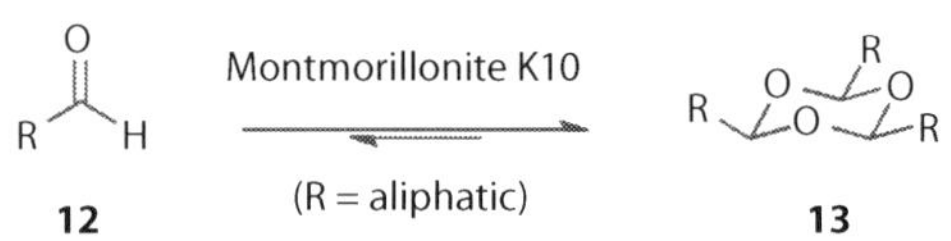

SCHEME 6.8 Clay-catalyzed cyclotrimerization of aliphatic aldehydes.

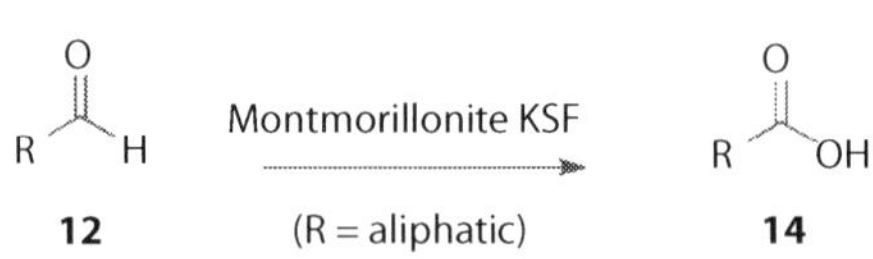

SCHEME 6.9 Montmorillonite KSF clay-catalyzed oxidation of aliphatic aldehydes.

TABLE 6.3
Scope of the Clay-Catalyzed Hetero-Diels–Alder Reaction

Entry	Aldehyde (12)	Product	% Yield
1	Propanal (**12a**)	**13a**	59
2	Butanal (**12b**)	**13b**	81
3	Isobutyraldehyde (**12c**)	**13c**	58
4	Pentanal (**12d**)	**13d**	95
5	Hexanal (**12e**)	**13e**	90
6	Cyclohexanecarboxyladehyde (**12f**)	**13f**	57

(Table 6.4) of homoallylic silyl ethers (**14**; Scheme 6.10) at room temperature or lower.[23]

Although other powerful Lewis acids ($SnCl_4$, $TiCl_4$, and $AlCl_3$) have been shown to successfully catalyze the Hosomi–Sakurai reaction,[24] our clay-catalyzed procedure comprises a marked improvement over existing methodologies with regard to sustainability: compared with most other metal halide-type Lewis acid catalysts, Montmorillonite K10 is dirt cheap (pun intended!), much easier to use, safer to handle, and essentially generates no waste. Additionally, because no aqueous workup is required, the clay-catalyzed reaction has the advantage of providing the products in their "protected" form, as trimethylsilyl ethers (rather than the homoallylic alcohols typically observed). This reaction, in fact, is one that proceeds with 100% atom economy and an E Factor of less than 1!

Another reaction we observed to proceed with near-perfect atom economy is the clay-catalyzed Prins–Friedel–Crafts-type synthesis of 4-aryltetrahydropyrans (**15**; Scheme 6.11). We modeled our study on Reddy and coworkers' reported one-pot multicomponent reaction of carbonyl compounds with 3-buten-1-ol in benzene, which was catalyzed by BF_3-OEt_2.[25] Our results (Table 6.5) were similar to those reported by

TABLE 6.4
Scope of the Clay-Catalyzed Hosomi–Sakurai Reaction

Entry	Aldehyde (9)	Product	% Conversion
1	Benzaldehyde (**9a**)	**14a**	75
2	*o*-Anisaldehyde (**9b**)	**14b**	70
3	*o*-Nitrobenzaldehyde (**9h**)	**14c**	84
4	*p*-Chlorobenzaldehyde (**9k**)	**14d**	99
5	*p*-Nitrobenzaldehyde (**9l**)	**14e**	99
6	*m*-Nitrobenzaldehyde (**9m**)	**14f**	99
7	*m*-Fluorobenzaldehyde (**9n**)	**14g**	97
8	2,6-Dichlorobenzaldehyde (**9o**)	**14h**	99
9	*m*-cyanobenzaldehyde (**9p**)	**14i**	99

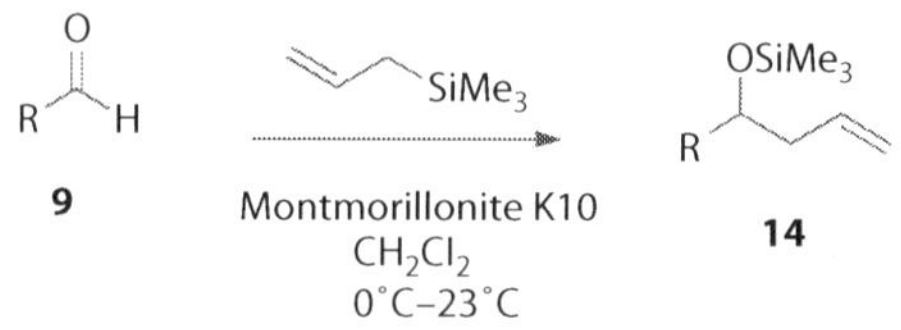

SCHEME 6.10 Clay-catalyzed Hosomi–Sakuraki-type reaction.

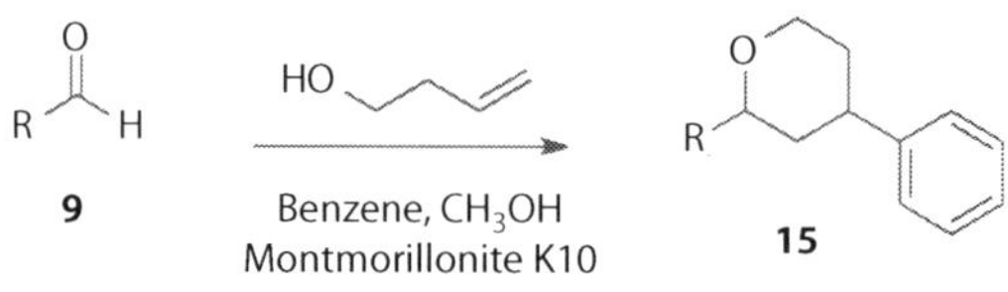

SCHEME 6.11 Clay-catalyzed Prins–Friedel–Crafts-type reaction.

TABLE 6.5
Scope of the Clay-Catalyzed Prins–Friedel–Crafts-Type Reaction

Entry	Aldehyde (9)	Product	% Conversion
1	*o*-Nitrobenzaldehyde (**9h**)	**15a**	99
2	*p*-Nitrobenzaldehyde (**9l**)	**15b**	90
3	*m*-Nitrobenzaldehyde (**9m**)	**15c**	88
4	*m*-Bromobenzaldehyde (**9q**)	**15d**	91

Reddy, although our approach is much more sustainable.[26] Not only is BF_3-OEt_2 considerably more expensive than Montmorillonite K10 clay, but it also must be handled under an inert environment, and reactions catalyzed by BF_3-OEt_2 typically require aqueous workup conditions that result in the generation of substantial waste products.

In keeping with the definition of sustainability, as originally described by the 1987 Brundtland Commission, Montmorillonite clays are ideally suited for moving chemistry into a direction of meeting "the needs of the present without compromising the ability of future generations to meet their own needs." With the ability and potential to catalyze organic reactions that are important for the synthesis of physiologically active compounds that are beneficial to society, at a substantially reduced cost, and with minimal impact on the environment, Montmorillonite clays are well on their way to meeting the Triple Bottom Line standard for new, sustainable technologies. Indeed, "the dark clay has a bright future in the area of organic synthesis."[6]

REFERENCES

1. Nagendrappa, G. *Appl. Clay Sci.* 2011, 53, 106–138.
2. Kaur, N., and Kishore, D. *J. Chem. Pharm. Res.* 2012, 4, 991–1015.
3. Anastas, P.T., and Levy, I.J. *Green Chemistry Education: Changing the Course of Chemistry*. American Chemical Society, Oxford University Press: New York, 2009.

4. Franchi, M., and Gallori, E. *Gene* 2005, 346, 205–214.
5. Plummer, C.C., McGeary, D., and Carlson, D.H. *Physical Geology*, 8th ed. McGraw-Hill Companies: Boston, 1999.
6. Nagendrappa, G. *Resonance* 2002, 7 (1), 64–77.
7. Mizuno, N., and Misono, M. *Chem. Rev.* 1998, 98, 199–218.
8. Dintzner, M.R., McClelland, K.M., and Coligado, D. Progress toward the synthesis of an anti-inflammatory acetophenone glucoside. Abstracts of Papers, *225th National Meeting of the American Chemical Society*, New Orleans, LA; American Chemical Society: Washington, DC; ORGN 429, 2003.
9. Dauben, W.G., Cogen, J.M., and Behar, V. *Tetrahedron Lett.* 1990, 31, 3241.
10. Dintzner, M.R., Morse, K.M., McClelland, K.M., and Coligado, D.M. *Tetrahedron Lett.* 2004, 45, 79–81.
11. Hepworth, J.D., Gabbutt, C.D., and Heron, M.B. *Comprehensive Heterocyclic Chemistry II*. Pergamon: New York, 1996, 301–350.
12. Dintzner, M.R., McClelland, K.M., Morse, K.M., and Akroush, M.H. *Synlett* 2004, 11, 2028–2030.
13. Brooks, G.T., Pratt, G.E., and Jennings, R.C. *Nature* 1979, 281, 570.
14. Cardillo, G., Cricchio, R., and Merlini, L. *Tetrahedron* 1968, 24, 4825.
15. Dintzner, M.R., Lyons, T.W., Akroush, M.H., Wucka, P., and Rzepka, A.T. *Synlett* 2005, 5, 785–788.
16. Dintzner, M.R., Wucka, P.R., and Lyons, T.W. *J. Chem. Educ.* 2006, 83, 270–272.
17. Chambers, J., Crawford, J., Williams, H.W.R., Dufresne, C., Scheigetz, J., Bernstein, M.A., and Lau, C.K. *Can. J. Chem.* 1992, 70, 1717.
18. Dintzner, M.R., Little, A.J., Pacilli, M., Pileggi, D.J., Soner, Z.R., and Lyons, T.W. *Tetrahedron Lett.* 2007, 48, 1577–1579.
19. Bednarski, M.D., and Lyssikatos, J.P. In: *Comprehensive Organic Synthesis*, edited by Trost, B.M., and Fleming, I. Pergamon Press: Oxford, UK, 1991; vol. 2, pp. 661–706.
20. Dintzner, M.R., Mondjinou, Y.A., and Pileggi, D.J. *Tetrahedron Lett.* 2010, 51, 826–827.
21. Larock, R.C. *Comprehensive Organic Transformations: A Guide to Functional Group Preparations*. VCH: New York, 1989, and references cited therein.
22. Hosomi, S., and Sakurai, H. *Tetrahedron Lett.* 1976, 16, 1295.
23. Dintzner, M.R., Mondjinou, Y.A., and Unger, B. *Tetrahedron Lett.* 2009, 50, 6639–6641.
24. Sasidharan, M., and Tatsumi, T. *Chem. Lett.* 2003, 32, 624.
25. Reddy, U.C., Bondalapati, S., and Saikia, A.K. *J. Org. Chem.* 2009, 74, 2605.
26. Dintzner, M.R. *Synlett* 2013, 24, 1091–1092.

7 Harnessing Solar Energy
Transition Metal Catalysts for the Water Oxidation Process

Margaret H. Roeder, Bryan C. Eigenbrodt, and Jared J. Paul

CONTENTS

7.1 MOTIVATION FOR THE DEVELOPMENT OF WATER OXIDATION CATALYSTS

One of the greatest challenges in the history of human civilization involves capturing and securing useful energy. Access to these energy sources can be related directly to the rise and fall of several of the world's most successful civilizations.[1] Over time, society has become reliant on fossil fuels to maintain its industrial economy with little use of nuclear power or renewable energy sources.[1] Successful advancements made during the Industrial Revolution launched a dependence on energy to power and maintain technological, economic, social, and political infrastructures.[1] With these industrial developments, energy supply remains an area of high concern along with the environmental impacts of obtaining useful energy sources.[1] The only solution to secure the world's energy future revolves around the development of a society based on clean, renewable energy.

Because society as a whole has become reliant on energy to maintain its progressive lifestyle, a shift to more efficient and economical energy sources must occur. Significant renewable energy sources include hydroelectric, geothermal, wind, tides, and solar energy.[2] However, solar energy remains the only renewable energy source with the capacity to provide enough power for the entire planet.[2] Harnessing solar energy provides a way to limit society's reliance on nonrenewable fossil fuels, which emit pollutants into the atmosphere. Some applications of solar energy include solar-powered cars and solar panels that convert energy from the sun into electricity.[3] In addition, transparent photovoltaic glass used in windows and doors convert sunlight into electricity.[4] Although utilizing the sun to produce electricity directly remains one specific aim of many researchers, this process only works when the sun shines. Finding ways to harness the sun's energy and store it as a fuel for future use remains the central goal in solar energy conversion. On a large scale, the production of fuels from solar energy via nanostructures or devices requires the generation of reductive equivalents (i.e., protons and electrons) for energy production. Storing these reductive equivalents in energy-rich chemical bonds can provide fuels that people can use at any time during the day or night (Figure 7.1). A major challenge in developing devices that produce fuel from the sun remains the source of these reductive equivalents.

Through photosynthesis, nature has found a way to convert energy from the sun into chemical energy, which provides fuel for plants. Nature uses sunlight as an energy source and water as the source of reductive equivalents, which power the electron transfer chain for the photosynthetic process. Photosynthesis as well as artificial photosynthetic devices require several key components, including "(1) light absorption, (2) excited state energy and electron transfer, (3) electron/proton transfer driven by free energy gradients, and (4) electron transfer activation of catalysts for oxidation and reduction."[5] Water oxidation yields four electrons that replenish the electrons transferred from Photosystem II to Photosystem I to generate adenosine triphosphate (ATP), which stores chemical energy. The generation of ATP through the photosynthetic process remains necessary for the conversion of carbon dioxide into carbohydrates, which provide plants with energy

$$2\,H^+ + 2\,e^- + \rightarrow H_2$$

$$CO_2 + 4\,H^+ + 4\,e^- \rightarrow CH_4 + O_2$$

$$2\,CO_2 + 8\,H^+ + 8\,e^- \rightarrow 2\,CH_3OH + O_2$$

FIGURE 7.1 Representative reactions showing the use of protons/electrons to create energy-rich chemical bonds that ultimately generate useful fuels.

for survival.[6] For artificial photosynthetic devices, water serves as the sole source of reductive equivalents necessary for solar fuel generation.[2] In photosynthesis, a manganese–calcium cluster catalyzes the water oxidation process due to the high energy associated with multistep reaction mechanisms. The design of catalysts that can efficiently carry out the water oxidation process is an area of intense research and is the focus of this chapter.

7.2 PHOTOSYNTHESIS AND PHOTOSYSTEM II

In photosynthesis, plants use energy from the sun, carbon dioxide, and water to make energy-rich carbohydrates for fuel, with oxygen being the by-product. The chemical reactions that convert carbon dioxide and water to carbohydrates and oxygen are reduction–oxidation reactions, consisting of two half-reactions (Figure 7.2). These reactions require multiple steps involving the transfer of electrons and protons while also avoiding high-energy intermediates. Therefore, proton-coupled electron transfer (PCET) reactions play an important role in these energy conversion and storage systems.[7] These reactions are endothermic, requiring energy in the form of sunlight. Water oxidation provides the electrons and protons necessary for the generation of ATP and nicotinamide adenine dinucleotide phosphate (NADPH). The NADPH produced donates electrons to carbon dioxide to produce sugars. As a result, these PCET reactions involve the ultimate transfer of protons and electrons from water to carbon dioxide to generate carbohydrates for plants.

The photosynthetic apparatus in natural photosynthesis acts like a wire, passing electrons from the site of water oxidation through the wire for use in carbon dioxide fixation, a process that converts carbon dioxide into useful carbohydrates (Figure 7.2; Equation 7.4). To "help" the electron along, photosynthesis has two light-absorbing components that provide the energy necessary to promote electron transfer, Photosystem I and Photosystem II. The following discussion will focus on Photosystem II, which contains the water oxidation catalyst.

Chlorophyll P680 serves as the central component of Photosystem II (Figure 7.3), aptly named because it absorbs the 680 nm wavelength of light.[6] This light absorption promotes an electron into a higher energy excited state in chlorophyll and primes Photosystem II for a series of electron transfer events. The high-energy electron is passed on to a nearby heme, pheophytin D_1, followed by electron transfer to a quinone, benzoquinone Q_A. The electron continues to be passed along, ultimately being used to generate reductive equivalents for carbon dioxide reduction. When the electron is promoted to the excited state in chlorophyll P680, an electron hole results, which needs to be filled to regenerate the ground state chlorophyll P680 for another photon to be absorbed and the process to continue. An electron hole refers to the

$$6\,CO_2 + 24\,e^- + 24\,H^+ \longrightarrow C_6H_{12}O_6$$

$$2\,H_2O \longrightarrow O_2 + 4\,e^- + 4\,H^+$$

FIGURE 7.2 Representative reactions of (4) carbon dioxide reduction and (5) water oxidation.

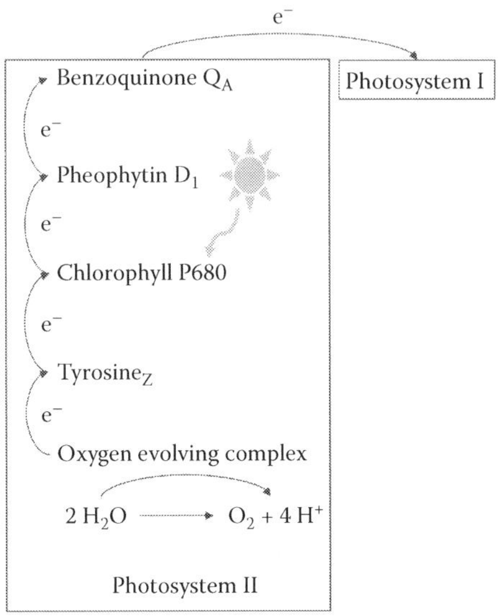

FIGURE 7.3 Schematic of the electron transfer chain found within Photosystem II.

vacancy in chlorophyll P680 once the absorption of a photon excites an electron to a higher energy state. The chlorophyll receives the electron from a nearby tyrosine amino acid, tyrosine Z. The oxidized tyrosine is subsequently reduced by the oxygen-evolving complex (OEC), the site of water oxidation. Ultimately, water serves as the sacrificial reductant fuel source, supplying electrons that are passed through a "wire" to keep the photosynthetic apparatus running while producing oxygen as the by-product.

The OEC (Figure 7.4), located in Photosystem II, catalyzes water oxidation in the photosynthetic process. The core metal ion–containing structure of the OEC (Figure 7.5) consists of four manganese ions, one calcium ion and five oxygen atoms that combine to form a Mn_4CaO_5 cluster.[8] Specifically, three manganese ions, four oxygen atoms, and the calcium ion form a cubane structure with the fourth manganese ion lying outside this structure linked by a di-oxo bridge.[8] In addition, four

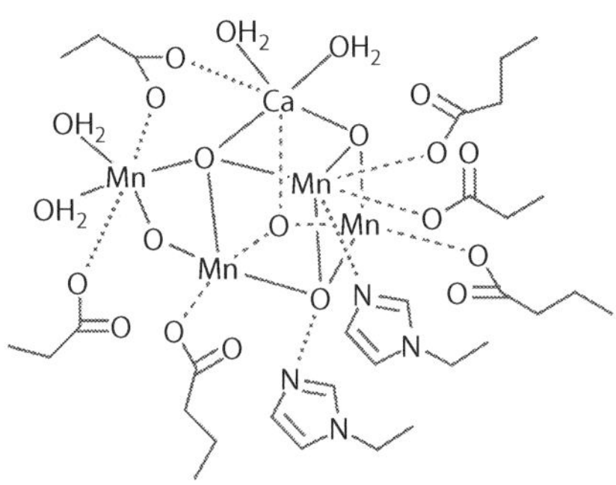

FIGURE 7.4 Structure of the OEC and surrounding amino acids found in Photosystem II.

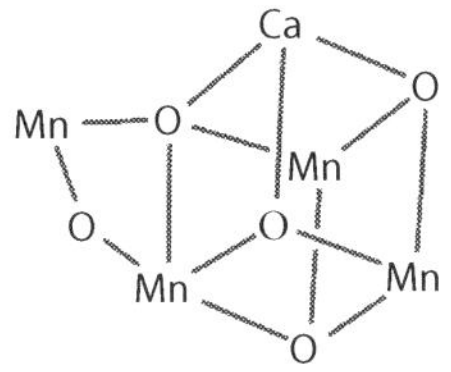

FIGURE 7.5 Structure of the OEC cubane comprised of an Mn_4CaO_5 cluster.

water molecules associated with two of the metal ions play an important role in the water oxidation process. Two water molecules attached to the manganese ion outside the cubane and two water molecules coordinated to the calcium ion possibly serve as substrates for water oxidation (Figure 7.4).[8] The coordination of water molecules to these metal centers ultimately generates the oxygen–oxygen bond for its release.

Amino acids that surround the OEC play a significant role in the water oxidation process. Some of these amino acids coordinate directly with the manganese and calcium ions whereas several others play a role in hydrogen bonding networks. These networks stabilize the cubane and aid in deprotonation of water molecules for the production of oxygen.[6,8] The manganese ions of the metal cluster remain necessary and play an important role in the release of oxygen. Through stepwise electron/proton loss, the manganese ions reach a higher oxidation state to bind water and eventually release oxygen. The calcium ion in the cubane structure also serves an important purpose in the water oxidation process. One proposed role of calcium involves binding substrate water molecules, allowing for the formation of oxygen.[9] Calcium may also play a role in controlling the transfer of protons as a way for the complex to avoid charge buildup and high-energy intermediates.[9]

The synthesis of complexes that replicate the structure and activity of the tetramanganese core of Photosystem II remains significant because of its high catalytic efficiency. These biomimics, which are functional models of this complex, provide insight into its mechanism of water oxidation as well as potentially generate practical water oxidation catalysts for artificial photosynthetic devices. In 2011, Kanady et al.[10] reported a synthesis of a $[Mn_3CaO_4]^{6+}$ cubane that matches the structure of the OEC without the fourth manganese ion outside the complex. The Mn^{IV} ions in the cubane possess the same oxidation states as the manganese centers in the OEC.

In addition to this complex, Kanady et al.[10] also synthesized a $[Mn_4O_4]^{6+}$ cubane structure to determine the properties of the calcium ion in the OEC. The comparison of these two synthesized models led to a better understanding of calcium's role in water oxidation as well as the overall mechanism of cubane formation. Cyclic voltammetry studies showed that the presence of calcium supported the formation of manganese ions with higher oxidation states at a lower potential.[10] Therefore, calcium remains vital to the OEC because it helps manganese access higher oxidation states, a significant step in the mechanism of photosynthetic water oxidation. Many other synthetic manganese clusters have been developed, including manganese dimers, manganese oxides, $CaMn_3O_x$ clusters, and manganese oxides that incorporate metals other than calcium.[11] These compounds "supplement the structural, spectroscopic, and mechanistic studies of biological systems" as well as provide potential catalysts for water oxidation.[11] Additionally, synthetic manganese clusters serve as biomimetic models of the OEC, providing insight into its mechanism of action.

7.3 WATER OXIDATION

The mechanism of oxygen evolution by the OEC in photosynthesis remains relevant in the development of water oxidation catalysts. The net conversion of water to oxygen is complicated by the fact that four electrons and four protons must be removed and an oxygen–oxygen bond must be formed (Figure 7.6; Equation 7.6). Although a photon of absorbed light possesses enough energy to carry out this overall four-electron/four-proton process, the individual steps remain significantly uphill in energy. These more fundamental one-electron/one-proton and two-electron/two-proton reactions provide significant energy barriers to the overall process. A photon associated with the absorption of 680 nm of light possesses 1.8 V and the reduction potential of pheophytin D_1 is −0.6 V, which leaves 1.2 V available for water oxidation.[12] Figure 7.6 shows standard reduction potentials of some possible water oxidation mechanisms in aqueous solution reported at pH 0 and calculated at pH 7.[13] The Nernst equation (Figure 7.7; Equation 7.9) relates the formal reduction potential (E°) to the half-wave potential at pH 0 ($E_{1/2}$), demonstrating the pH dependence of reduction potentials.[14] The calculated values at pH 7 remain more relevant to the natural photosynthetic process because physiological pH is approximately 7. The energetics associated with these processes are critical in developing catalysts for water oxidation.

The exact mechanism of water oxidation remains unknown, but several theories on the process and the role of the OEC have been proposed.[6,9,15,16] The OEC exists in five intermediate states (S_0–S_4), which leads to the formation of dioxygen.[6] The absorption of four photons of light in sequential steps drives the oxidation of two water molecules as the OEC progresses through the five states, known as the Kok cycle.[16] When the manganese ion outside of the cubane structure of the OEC reaches a high oxidation state, dioxygen is formed and the uptake of two more water molecules reduces the complex again.

Possible water oxidation reaction mechanisms	Reduction potential (E°) at pH 0	Reduction potential (E°) at pH 7
(6) $2\,H_2O \rightarrow O_2 + 4\,H^+ + 4\,e^-$	1.229	0.815
(7) $H_2O \rightarrow \bullet OH + H^+ + e^-$	2.848	2.435
(8) $2\,H_2O \rightarrow H_2O_2 + 2\,H^+ + 2\,e^-$	1.776	1.35

FIGURE 7.6 Representative equations of possible water oxidation reaction mechanisms and corresponding E° values at pH 0 and 7: (6) four-electron reaction yielding oxygen and four protons (7) one-electron reaction yielding a hydroxyl radical and one protons (8) two-electron reaction yielding hydrogen peroxide and two protons.

$$E^{\circ} = E_{1/2} - \frac{0.05916}{n}(\text{pH})$$

FIGURE 7.7 (9) Nernst equation demonstrating the pH dependence of reduction potential where E° = formal reduction potential, $E_{1/2}$ = half-wave potential at pH 0, n = number of electrons.

One proposed mechanism (Figure 7.8) involves the nucleophilic attack of water on a high oxidation state Mn-oxo group.[15] Initially, a water molecule attacks the manganese ion outside of the Mn_3O_4Ca cluster and forms an electron-deficient metal-oxo group after several deprotonation steps in the Kok cycle.[6] A second water molecule coordinates to a manganese or calcium ion within the oxygen-evolving cluster. The nucleophilic attack of this water molecule or hydroxide group on the metal-oxo group leads to the generation of a peroxide intermediate, which then forms dioxygen.[16]

A second proposed mechanism (Figure 7.9) involves the formation of an oxyl radical (Mn=O•) after a water molecule coordinated to the manganese ion outside the cluster undergoes several deprotonation steps.[6] This radical can attack a second water molecule coordinated to a manganese or calcium ion within the cubane structure to form dioxygen. Alternatively, the oxyl radical can coordinate with one of the oxygen atoms within the Mn_3O_4Ca cluster, which also leads to the formation of dioxygen.[6] The development of an OEC mimic remains essential to studying its structure and mechanism of action. Based on the proposed mechanisms, the ability of manganese to reach high oxidation states and bind water is a key feature for water oxidation.

The manganese–calcium cluster in the OEC plays a key role in the water oxidation process, so catalysts for artificial photosynthetic devices must include transition metals. More than 30 elements make up the transition metals on the periodic table, leading to a wide variety of systems. Due to the progressive nature of catalyst development, discussing each and every water oxidation catalyst remains impractical, so this chapter will instead highlight several examples based on transition metals.[2,15,17–20] The introductory section of this review will discuss catalyst terminology, including homogeneous and heterogeneous catalysis, measurements of catalysis, and

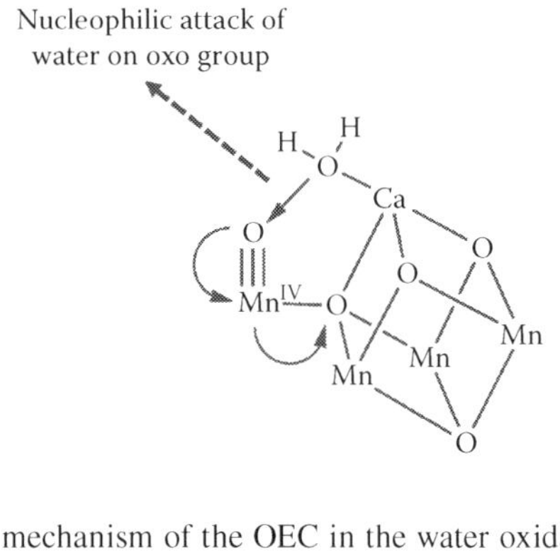

FIGURE 7.8 Proposed mechanism of the OEC in the water oxidation process involving a nucleophilic attack of water on Mn≡O oxo group.

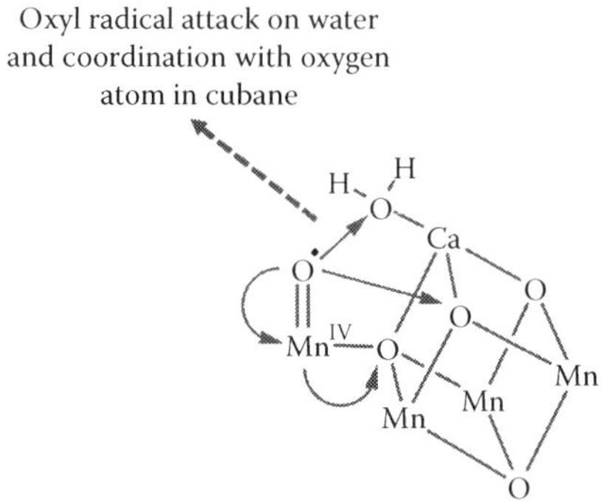

FIGURE 7.9 Proposed mechanism of the OEC in the water oxidation process involving the attack of an oxyl radical (Mn=O•) on water and coordination of the oxyl radical with an oxygen atom within the OEC cubane.

terms relating to the activation of catalysts. Overall, the main focus of this chapter revolves around the use of transition metals in the development of water oxidation catalysts that possess the potential for solar energy conversion and artificial photosynthetic devices.

7.4 TERMINOLOGY FOR CATALYSIS IN WATER OXIDATION

7.4.1 Standard Catalysis Terms

Three parameters assess the activity and effectiveness of water oxidation catalysts: turnover number (TON), turnover frequency (TOF), and overpotential.[15] The TON of a catalyst refers to the number of moles of substrate that a catalyst converts to product before deactivation (Figure 7.10; Equation 7.10).[2] Because the TON exists as a ratio between the moles of product formed per mole of catalyst, it remains unitless. For example, a TON of 100 correlates to 100 moles of oxygen per catalytic site. A large-scale artificial photosynthetic device should last for approximately 10^8 turnovers to remain economically practical.[2] A significant requirement of water oxidation catalysts involves a high TON, which correlates to a greater production of oxygen. The longer the lifetime of a catalyst, the higher the TON as long as the catalyst remains intact. Therefore, the TON is an indication of a catalyst's ability to produce molecules of oxygen, longevity, and stability against degradation.

The TOF measures the number of cycles a catalyst undergoes or the TON per unit time (Figure 7.11; Equation 7.11).[2,15] Typically, units for TOF are seconds^{-1} or minutes^{-1}. An ideal water oxidation catalyst would possess a TOF greater than the rate that photons flow from the sun.[15] A high TOF results in a high quantum yield, which refers to the electrons generated from the absorption of photons.[15] Similar to the OEC

$$\text{Turnover number (TON)} = \frac{\text{Moles of product formed}}{\text{Moles of catalyst}}$$

FIGURE 7.10 (10) Representative equation of TON.

$$\text{TOF} = \frac{\text{TON}}{\text{Unit time}} \quad (11)$$

FIGURE 7.11 (11) Representative equation of TOF.

in Photosystem II, artificial photosynthetic devices require rapid electron transfer from the catalyst to compete with high back electron transfer rates.[15] Fast back electron transfer will prevent the catalyst from turning over as it refills the electron hole and puts the catalyst back in its resting state. In Photosystem II, back electron transfer between benzoquinone Q_A and chlorophyll P680 occurs in the inverted region in which electron transfer rates decrease as the driving force for the reaction increases.[5] Therefore, water oxidation requires catalysts with high TOFs to function efficiently.

Overpotential refers to "the additional driving force needed to overcome the activation energy of the rate-limiting step" for a reaction to proceed.[2] As the catalyst operates, this excess energy is necessary to push the water oxidation process forward at a faster pace for better efficiency.[15] In electrochemistry, overpotential signifies "the difference between the applied potential where catalysis begins and the thermodynamic potential."[2] Catalysts that increase the rate of a reaction with little or no overpotential are the most desirable for water oxidation because they minimize the energy requirements for catalysis. However, the water oxidation reaction itself possesses a high overpotential due to multiple steps that form high-energy intermediates, which makes the development of a catalyst difficult.[15,18] Decreasing the overpotential of catalysts for the water oxidation process remains significant in the development of an efficient artificial photosynthetic device.

Before a catalyst can function, it must undergo an activation process that brings it to a state in which it can oxidize water. Catalyst activation occurs chiefly through three different routes: photochemical activation, the use of sacrificial oxidants, and direct electrochemical oxidation. Photochemical activation of catalysts involves the use of a photosensitizer (PS), a compound that absorbs a photon and is promoted to an excited state (PS*).[21] Ideally, then, the catalyst can quench the excited state of the photosensitizer through electron transfer, thus activating the catalyst to begin the water oxidation process (Figure 7.12).[21] Conjugated organic molecules as well as metal complexes capable of metal-to-ligand charge transfer (MLCT) can be used as photosensitizers due to their light-absorbing properties.[21] Ruthenium polypyridyl complexes, specifically $[Ru(bpy)_3]^{2+}$ (bpy = 2,2′-bipyridine), remain some of the

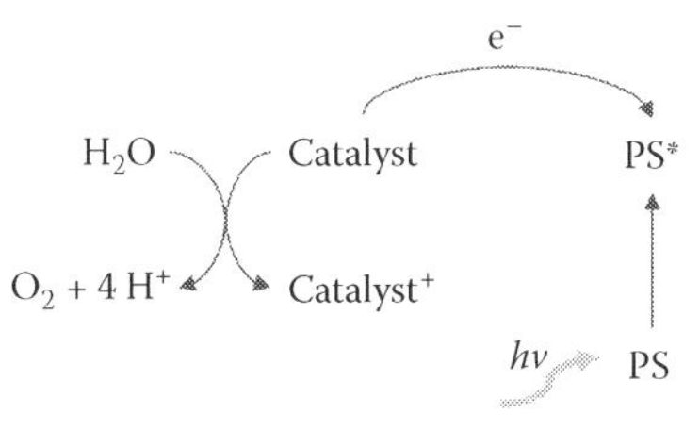

FIGURE 7.12 Schematic of photochemical activation of a catalyst involving the absorption of a photon by a photosensitizer and the quenching of its excited state through electron transfer from the catalyst where PS = photosensitizer, PS* = excited state of photosensitizer.

most well-known complexes in light absorption because of their chemical stability, redox properties, and excited state activity.[21,22] Photosensitizers are a desirable way to activate catalysts because they couple light energy absorbance to the water oxidation process, similar to water oxidation in photosynthesis.

In contrast to photochemical catalyst activation, electrochemical techniques involve electrolysis controlled by potential.[23] In this case, catalysts are typically attached to an electrode surface and connected to a potential source to oxidize the metal center. Charge transfer between the electrode and the metal center occurs during electrolysis. Homogeneous catalysts in solution typically use the electrochemical method, but the complex must adhere to an electrode surface to function.[23] Because electrochemical catalyst activation requires an external source of energy, this method is nonideal.

Lastly, catalysis may require a sacrificial oxidant, which directly oxidizes the metal center so it can react with water to form oxygen.[18] Chemical oxidants, such as persulfate, cobalt(III), and cerium(IV), possess the strong oxidative potentials necessary to activate the catalyst and drive water oxidation.[15,21] A disadvantage of chemical oxidants involves the generation of highly oxidizing species that lead to various side reactions.[15] Sacrificial oxidants provide a better understanding of water oxidation catalysis, but photochemical catalyst activation through light absorption remains the most favorable method.

7.4.2 Heterogeneous Catalysis

Heterogeneous catalysis involves the use of nanoparticles, films, or other insoluble species to catalyze water oxidation. Heterogeneous catalysts include colloids supported on a substrate or surface, nanoparticles, and metal oxides based on transition metals, such as manganese, cobalt, iron, ruthenium, or iridium.[15] Properties associated with heterogeneous catalysts include low cost, ease of interface with electrode systems, and oxidative stability.[24] Some advantages of heterogeneous catalysts include stability, high activity, and longevity. These catalysts tend to be more robust, more easily prepared, and less expensive than homogeneous catalysts.[2] An important aspect of heterogeneous catalysts involves their ability to be easily separated via filtration.[25] Therefore, heterogeneous catalysts can be used more than once, an attractive quality for wide-scale solar energy conversion. Disadvantages of heterogeneous catalysis include difficulty in mechanistic studies, deactivation of catalysts through clumping, and the variation in surface sites to which the catalysts attach. Figure 7.13

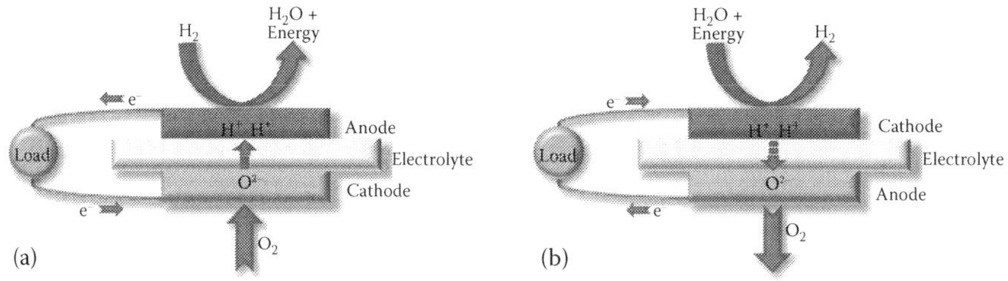

FIGURE 7.13 Schematic of different heterogeneous systems: (a) fuel cell system, running at a negative potential bias; and (b) electrolysis system, running at a positive potential bias.

depicts a schematic of a general heterogeneous catalyst system in the electrolysis of H_2O. Depending on the voltage bias that is applied, this system can perform as a fuel cell (negative potentials) as well as a water electrolyzer (positive potential).

7.4.3 Homogeneous Catalysis

Homogeneous catalysis (Figure 7.14) involves soluble, molecular species with the capacity to catalyze water oxidation. The ability to investigate and characterize these complexes makes their development and improvement more easily attainable.[20] The information gained from these studies may lead to greater turnover rates, the ability to be combined with photosensitizers, and better stability.[2] Molecular catalysts typically possess higher catalytic activity, but remain less stable than heterogeneous catalysts during long-term catalysis.[18] These catalysts possess high turnover frequencies because a soluble species possesses a larger surface area than a particle or film.[20] A disadvantage of homogeneous catalysts involves thermodynamic instability due to organic ligands that are prone to oxidative degradation. Two examples of homogeneous catalysts include polyoxometalates (POMs) and molecular complexes studied in solution.

Although homogeneous and heterogeneous catalysts possess advantages that makes them desirable for implementation in solar energy conversion, three features remain essential for all water oxidation catalysts: speed, selectivity, and stability.[2] The stability of a catalyst remains a significant factor in its ability to catalyze water oxidation because it must remain active for a certain period to function. The catalyst must also be able to accumulate four oxidizing equivalents without charge buildup.[2] Similar to the OEC in Photosystem II, water oxidation catalysts avoid charge buildup through PCET processes, in which protons and electrons transfer together.

In addition to speed, selectivity, and stability, one of the most important requirements of water oxidation catalysts involves the cost of production. Considering the cost of developing catalysts for large-scale water oxidation and solar fuel production, the materials used to create an artificial photosynthetic device must be inexpensive.[2]

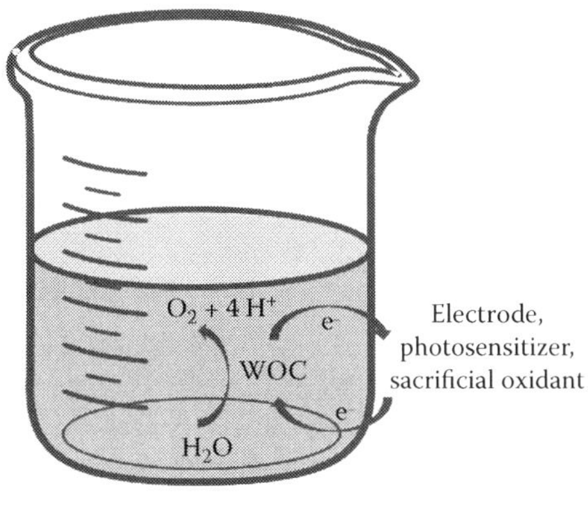

FIGURE 7.14 Schematic of a homogeneous system comprised of a soluble, molecular water oxidation catalyst (WOC) activated via electron transfer with an electrode, photosensitizer, or sacrificial oxidant.

The development of a biomimetic model of the OEC requires expansive research efforts to test the catalyst and make necessary improvements. From a logistical, economical, and environmental standpoint, the use of earth-abundant starting materials in the synthesis of water oxidation catalysts remains essential to solar fuel production on a global scale. Nature utilizes earth-abundant materials (i.e., manganese and calcium) for its water oxidation process in photosynthesis, so the use of expensive and rare elements seems wasteful and undesirable for practical applications. Therefore, a shift to the use of earth-abundant first-row transition metals remains critical to the advancement of harnessing solar energy.

7.5 THE TRANSITION METALS

Synthetic water oxidation catalysts incorporate transition metals (Figure 7.15) due to their ability to adopt multiple oxidation states (needed to mediate multiple electron transfers) as well as to bind ligands to form complexes (that can be covalently modified to precisely tune their properties). Second- and third-row transition metals, such as ruthenium and iridium, remain more expensive and rarer than first-row transition metals.[20] However, complexes based on earth-abundant, first-row transition metals (e.g., cobalt, copper, iron, manganese, and nickel) tend to exhibit lower catalytic activity than the second- and third-row transition metals for water oxidation.[20] The advancement of water oxidation catalysts comprised of inexpensive materials capable of catalysis under simple conditions remains a significant factor in discovering solar energy conversion and storage applications.

The next portion of this chapter involves a survey of different transition metals that serve as water oxidation catalysts. Each section begins with background information about a specific metal and some of its properties that make it attractive for water oxidation catalysis. These desirable qualities contribute to catalyst design with the ultimate goal being tunable properties, the ability to turn the catalyst on/off easily, and light absorption capabilities. In addition, the individual sections contain specific examples of transition metal complexes that catalyze water oxidation. This is by no means an exhaustive list and it is not intended to be. The first two sections discuss second- and third-row transition metals belonging to the platinum group

Transition metals

24 Cr	25 Mn	26 Fe	27 Co	28 Ni	29 Cu	30 Zn
42 Mo	43 Tc	44 Ru	45 Rh	46 Pd	47 Ag	48 Cd
74 W	75 Re	76 Os	77 Ir	78 Pt	79 Au	80 Hg

FIGURE 7.15 Schematic of earth-abundant first-row transition metals (dark gray) and rare second/third-row transition metals (light gray) used in water oxidation catalysis and discussed in this work.

metals, which consist of scarce and expensive metals. These metals have been widely used as water oxidation catalysts due to their high catalytic activity. In recent years, a shift to first-row transition metals has taken precedence over the platinum group metals due to their potential for practical applications. The remaining sections examine first-row transition metals that remain less expensive and more earth-abundant than second/third-row transition metals.

7.5.1 Ruthenium

Ruthenium is a second-row transition metal that belongs to a subclass known as the platinum group metals. These metals possess high catalytic activity for water oxidation, but remain scarce and more expensive than earth-abundant first-row transition metals.[20] Specifically, ruthenium remains one of the most studied transition metals for solar energy catalysis.[22,26] The reasons for continued research in this area are numerous, including a heavy synthetic precedence and the stability of transition metal complexes based on ruthenium. Therefore, a wealth of variations on the ligands that coordinate with ruthenium can lead to the synthesis of a limitless number of stable complexes.[16] Many ruthenium complexes absorb light in the visible region, taking advantage of the entire solar spectrum and giving promise for direct coupling of light absorbance to water oxidation.[22] These complexes often have long-lived excited states, resulting in slow charge recombination. Charge recombination regenerates the inactive ground state of the complex, which exists as a major hurdle to carrying out electron transfer chemistry. In addition, several ruthenium complexes are strongly luminescent, which makes exploring the excited state chemistry of these complexes relatively easy.[22]

Most notably, ruthenium complexes containing polypyridyl ligands possess a rich history. These complexes adopt an octahedral configuration and typically the most stable charge on the metal is Ru^{II}, which puts the ruthenium metal center in the d^6 electron configuration.[22] The polypyridyl ligands have low-lying π^* molecular orbitals, providing low-energy electronic transitions in the visible region of the MLCT type.[22] The light absorption properties of these complexes mean that they are prime candidates to have the light absorption coupled to the water oxidation process. In addition, higher oxidation states of ruthenium can be achieved including Ru^{III}, Ru^{IV}, and Ru^{V}, resulting in the potential for multielectron catalysis for reactions such as water oxidation.

The ruthenium complex, *cis,cis*-$[(bpy)_2(H_2O)Ru^{III}ORu^{III}(OH_2)(bpy)_2]^{4+}$, was the first molecular water oxidation catalyst (Figure 7.16).[26] The dimer consists of two ruthenium atoms bridged by an oxo linker. Each ruthenium atom has two bipyridine ligands located in the cis position relative to each other. A water molecule occupies the sixth coordination site. The lowest energy λ_{max} is pH dependent and appears between approximately 625 nm and approximately 642 nm, giving the complex its characteristic blue color for which it has been famously called the "blue dimer."[27] The structure of the blue dimer fixes two water molecules within proximity to each other (4.725 Å), which was hypothesized to prime the compound for oxygen–oxygen bond formation.[27] The coupling of two oxygen atoms is a major step in the water oxidation process.[28] The catalyst has achieved a TON of 13.2 and TOF of 0.0042 s^{-1} in 0.1 M $HClO_4$ at strongly acidic pH values.[29,30]

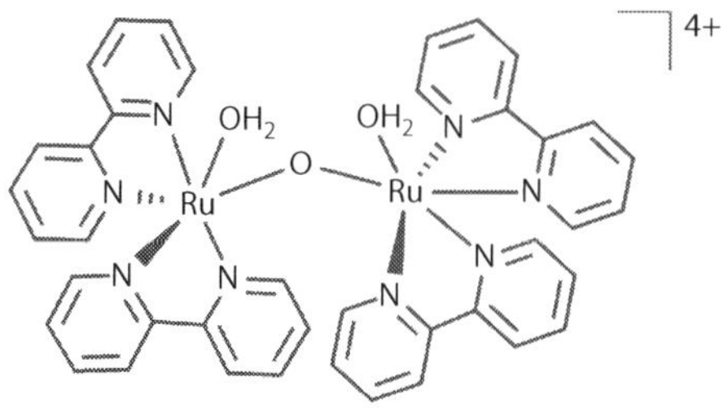

FIGURE 7.16 Structure of the first molecular water oxidation catalyst (the blue dimer) consisting of two ruthenium metal centers, four bipyridine ligands, an oxo linker between the two metal centers, and water molecules coordinated to each ruthenium center.

Several mechanistic studies have been carried out on the blue dimer and it has been determined that the $[(bpy)_2(O{=})Ru^{V}ORu^{V}({=}O)(bpy)_2]^{4+}$ oxidation state triggers the oxidation of water.[28] Once this oxidation state is reached, a water molecule attacks the $Ru^{V}{=}O$, which forms the peroxy intermediate Ru^{IV}-OOH.[31,32] These studies, most notably that an external water molecule not bound to the catalyst can for the oxygen–oxygen bond, have shed light on the fact that two aqua sites and the dimeric nature of the complex are not necessary for water oxidation catalysis.

A major advancement in the field of water oxidation catalysis occurred with the discovery that the blue dimer structure consisting of two water molecules was not necessary for the generation of oxygen, and only a single-site water-binding ruthenium complex could result in the oxidation of water.[33,34] In two highly investigated complexes,

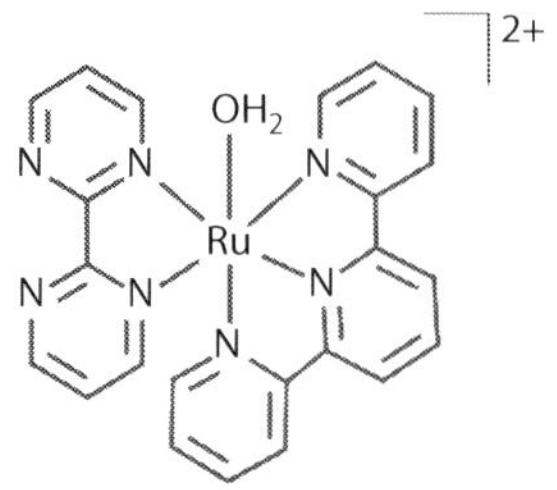

FIGURE 7.17 Structure of $[Ru(tpy)(bpm)(H_2O)]^{2+}$ where tpy = 2,2′:6′2″-terpyridine and bpm = 2,2′-bipyrimidine.

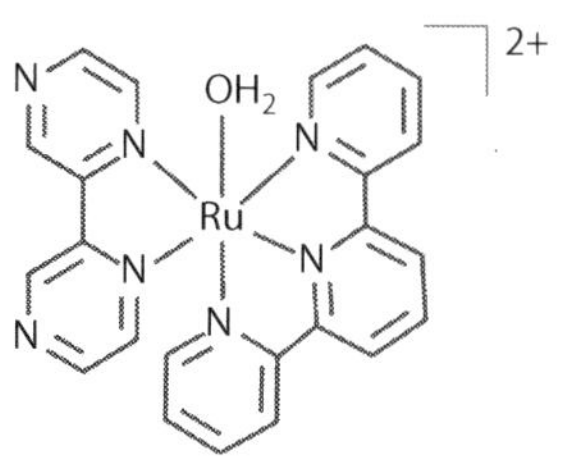

FIGURE 7.18 Structure of $[Ru(tpy)(bpz)(H_2O)]^{2+}$ where tpy = 2,2′:6′2″-terpyridine and bpz = 2,2′-bipyrazine.

$[Ru(tpy)(bpm)(H_2O)]^{2+}$ (Figure 7.17) and $[Ru(tpy)(bpz)(H_2O)]^{2+}$ (Figure 7.18) (bpm = 2,2′-bipyrimidine, bpz = 2,2′-bipyrazine, tpy = 2,2′:6′2″-terpyridine), catalytic water oxidation occurs when the Ru^{V}=O oxidation state is reached, similar to that observed with the blue dimer.[33] These results have led to a wealth of new single-site ruthenium catalysts with TONs as high as 3,200 and TOFs up to 0.018 s^{-1}.[34–40]

7.5.2 Iridium

Iridium, like ruthenium, belongs to the expensive and rare platinum group metals, so the cost of implementing an iridium-based catalyst into a large-scale solar energy conversion system remains more expensive than using a first-row transition metal. However, Ir^{III} d^6 complexes have been studied for their potential as water oxidation catalysts. An early complex found to oxidize water, $[(ppy)_2(H_2O)_2Ir^{III}]^+$ (Figure 7.19) (ppy = 2-phenylpyridine), possesses a TON of 2,490; the substituting of groups in the 5- and 5′-position on the ppy ligand only modestly affect the catalyst TON.[41] Further advancements arose with the discovery of Ir(Cp*) complexes (Figure 7.20) (Cp* = pentamethylcyclopentadiene), which possess TONs greater than an order of magnitude higher than the ppy complexes.[42] The mechanism of Ir(Cp*) complexes remain under investigation, but evidence exists for homogeneous catalysis as well as heterogeneous catalysis from nanoparticle formation.[43,44]

One goal in catalyst design involves the ability to turn the catalysts on and off by relatively simple changes in catalyst environment. Being able to reversibly turn a catalyst on and off in a switchable manner allows for greater control in that the reaction of interest will only turn over when desired. Designing catalysts with ligands that can adopt different protonation states exists as one way to significantly affect catalyst activity.[45] The ligands directly attached to the metal will possess different electronic effects on the metal center depending on the protonation state.[46–49] The additional bidentate ligand that can complex with the Ir(Cp*) scaffold serves as a basis for several studies. One class of compounds containing hydroxyl-substituted bipyridine ligands bound to iridium have been studied for their effects on catalytic water oxidation.[50,51] The ligands, 66′bpy$(OH)_2$ and 44′bpy$(OH)_2$ (4,4′-dihydroxy-bipyridine) increase electron-donating effects to a metal center upon deprotonation, readily observed in the resonance structures of the ligand (Figure 7.21). In addition, the 66′bpy$(OH)_2$ ligand has the potential to directly participate in water oxidation

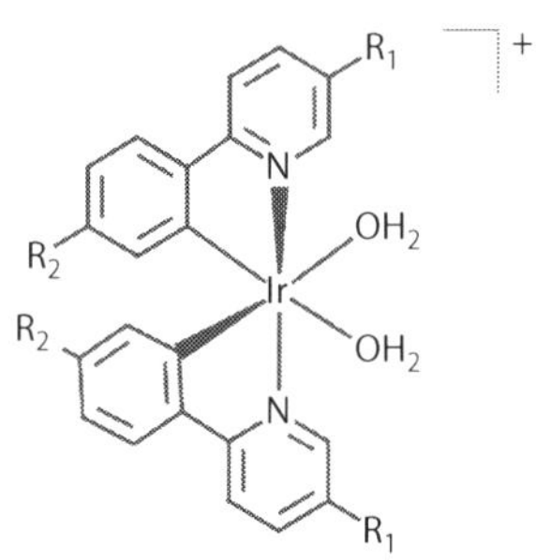

FIGURE 7.19 Structure of $[Ir^{III}(ppy)_2(H_2O)_2]^+$ where ppy = 2-phenylpyridine.

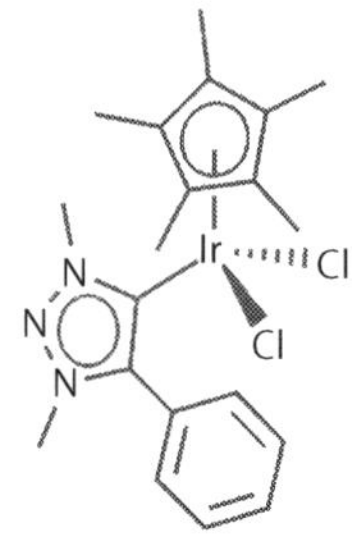

FIGURE 7.20 Structure of $[IrCl_2(Cp^*)(trz)]$ where Cp* = pentamethylcyclopentadiene and trz = triazolylidene.

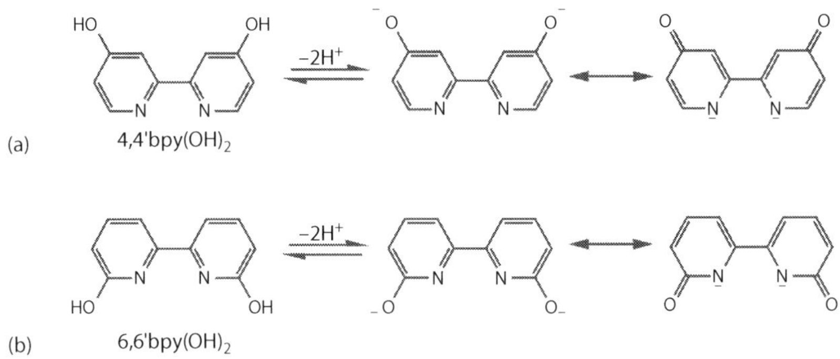

FIGURE 7.21 Schematic of resonance structures of (a) 4,4′bpy$(OH)_2$ and (b) 6,6′bpy$(OH)_2$ where bpy$(OH)_2$ = dihydroxy-bipyridine showing the electron-donating effects of the ligands upon deprotonation.

reaction with the hydroxyl group located close to the metal center, observed in the work with copper described previously.

Examining the catalytic activity of the complexes with 66′bpy$(OH)_2$ (Figure 7.22) and 44′bpy$(OH)_2$ ligands (Figure 7.23), the iridium catalysts shut off to water oxidation at low pH, whereas their catalyst activity turns on as the pH increases.[50] These results demonstrate the potential to not only tune a metal complex but also turn

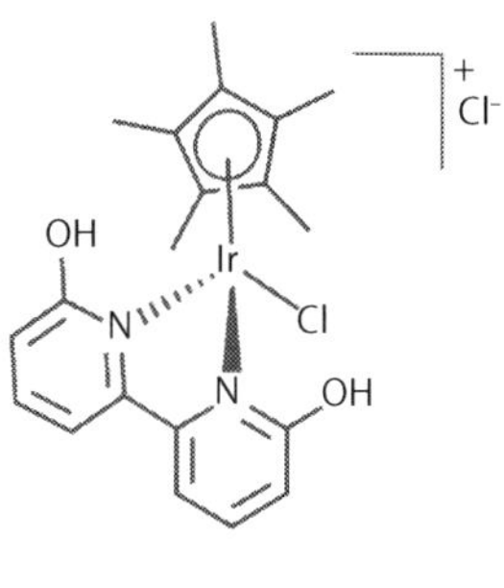

FIGURE 7.22 Structure of $[IrCl(Cp^*)(6,6'bpy(OH)_2]^+$ where Cp* = pentamethylcyclopentadiene and 6,6′bpy$(OH)_2$ = 6,6′-dihydroxy-bipyridine.

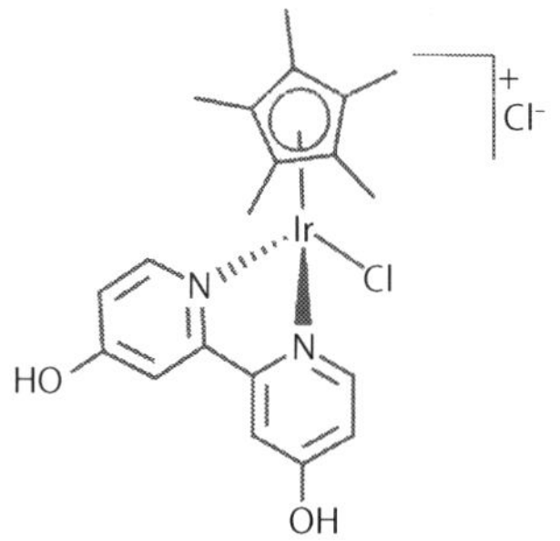

FIGURE 7.23 Structure of $[IrCl(Cp^*)(4,4'bpy(OH)_2]^+$ where Cp* = pentamethylcyclopentadiene and $4,4'bpy(OH)_2$ = 4,4′-dihydroxy-bipyridine.

catalysts into molecular on/off switches. The most effective catalysts studied possess a TON of 120 and a TOF of 0.2 s^{-1} at pH = 5.6 with the $44'bpy(OH)_2$ ligand compared with substitution in the $66'bpy(OH)_2$ position, which had a TON of 100 and a TOF of 0.17 s^{-1} at pH = 5.6.[50] These results indicate that orientation of the hydroxyl group toward the metal center possibly hinders reactivity toward water oxidation, but the influence of this ligand system remains under study.

7.5.3 Cobalt

Cobalt is a first-row transition metal that remains much more cost-efficient than precious metals (i.e., ruthenium, platinum, and iridium) for use in large-scale solar energy conversion systems. The metal mainly exists as Co^{II} and Co^{III} with d^7 and d^6 electron configurations, respectively.[52] Therefore, cobalt is electron rich, which makes it relevant for water oxidation catalysis. Similar to the OEC, cobalt exists in multiple oxidation states with redox potential values near those necessary to thermodynamically oxidize water.[53] Another advantage of using cobalt in the development of water oxidation catalysts involves its incorporation into metal-oxo frameworks. Cobalt can form stable metal-oxo structures in oxidizing environments like the Mn_4CaO_5 structure, which remain significant for the generation of catalysts that mimic the OEC.[53]

A disadvantage of using cobalt in photosynthetic devices involves its low earth abundance compared with iron, manganese, and nickel.[54] However, cobalt remains interesting for water oxidation catalysis because of its versatility and high catalytic activity. The Co^{II}, Co^{III}, and Co^{IV} oxidation states are accessible for the water oxidation process, but each require different stabilizing ligands.[55] Ligands that create a stable environment around the Co^{II} metal center differ from those that stabilize a Co^{III} center, which requires excess potential for oxidation.[55] Cobalt also remains attractive for water oxidation catalysis because it functions as both a molecular homogeneous catalyst as well as a heterogeneous catalyst in the form of a metal oxide.

The development of cobalt-based water oxidation catalysts from inorganic cobalt salts has emerged over the past few years. A cobalt phosphate (Co-P_i) catalyst that forms as a thin film on conducting surfaces serves as an example of a heterogeneous catalyst.[56] The oxidation of Co^{II} to Co^{III} results in self-assembly of the catalyst from aqueous solution, depositing onto the electrode surface as an oxide.[55] Co-P_i possesses the ability to work at neutral pH and deposit on several different electrode surfaces.[55] The catalyst functions while directly attached to semiconducting metal oxides, which provides it with an application for photoelectrochemical cells.

The Co-P_i catalyst also remains significant to water oxidation because it serves as a functional model of the OEC in Photosystem II. Both the OEC and the Co-P_i catalyst consist of earth-abundant first-row transition metals with the metal centers linked through bonds to oxygen atoms.[53] In addition, both catalysts self-assemble from water when the metal centers become oxidized. Both catalysts also possess a self-repair mechanism, which allows each complex to rebuild itself during catalysis.[55] This self-healing capability allows the complex to repair any damage that could break down the catalyst film during catalysis making them more robust. The manganese cluster in Photosystem II also possesses its own self-healing mechanism, which involves the reaction center D_1 protein in which the OEC resides.[55] The self-healing nature of catalysts remains significant to new advances in solar energy conversion because it allows a complex to perform multiple catalytic cycles before breaking down.

Another class of cobalt-based catalysts includes POMs that incorporate aspects of homogeneous and heterogeneous catalysis. Typically, POMs consist of "transition metal oxygen anion clusters" using anions, such as $[W_{10}O_{32}]^{4-}$ or $[Mo_7O_{24}]^{6-}$.[2] The homogeneous catalyst $[Co_4(H_2O)_2(PW_9O_{34})_2]^{10-}$ consists of a Co_4O_4 core stabilized by polytungstate ligands.[24] The polytungstate ligands of Co-POM possess oxidative resistance unlike carbon-containing ligands, which provides the catalyst with stability.[24] The catalyst requires the use of an oxidant to attain high oxidation states of cobalt so it can perform water oxidation. Similar to the Co-P_i catalyst, the Co-POM self-assembles in water from readily available cobalt, phosphorous, and tungsten salts.[24] However, Co-POM remains faster than Co-P_i with a TOF of 5 s^{-1} most likely due to the formation of soluble particles that possess a higher density of catalytically active cobalt sites.[53]

7.5.4 Copper

Copper serves as a cheap and abundant material for the development of new water oxidation catalysts. Copper possesses a history of binding and activating dioxygen, including proteins that transport oxygen as well as catalyze oxidation–reduction reactions, which makes the metal a suitable candidate for water oxidation catalysis.[57] Catalysts based on copper remain attractive because of their ability to form simple systems, high abundance of the metal, and low cost.[58] These metal complexes function with relatively high stability, but generating high oxidation state copper species, such as Cu^{III} or Cu^{IV}, can lead to high overpotentials for water oxidation.[58] These large overpotentials mean that the catalyst requires extra energy to reach high oxidation states and bind water. Due to this difficulty, copper may not serve as the best transition metal in the development of large-scale photosynthetic devices.

A potential advantage to using copper in catalysis involves its well-known coordination chemistry and the effectiveness of Cu^{I} and Cu^{II} in binding to organic molecules.[52] A main use of copper involves electron transfer systems that consist of oxidative enzymes and focus on energy capture.[52] Therefore, copper serves as a viable option for catalysis because redox chemistry and electron transfer remain crucial components in the water oxidation process. Copper also possesses broad redox chemistry with the ability to be reduced to Cu^0 and Cu^I as well as to be oxidized to Cu^{III} and Cu^{IV}, although the high overpotentials mentioned previously must be overcome.[59] In recent years, research in the field of water oxidation catalysis has focused on first-row transition metals, but copper has been explored less than other metals.

The first copper-based catalyst for electrolytic water oxidation involves a copper-bipyridine-hydroxo complex (Figure 7.24) that forms *in situ* from commercially available copper salts and bipyridine at pH values ranging from 11.8 to 13.3.[60] The catalyst acts a soluble molecular complex with $[(bpy)Cu(OH)_2]^{2+}$ as the dominant species.[60] The catalyst functions as one of the fastest homogeneous water oxidation catalysts with a TOF of 100 s^{-1}.[60] Most of the catalyst remains present after 30 turnovers, which means that the complex is relatively robust.[60] The mechanism of the $[(bpy)Cu(OH)_2]^{2+}$ remains under study to fully understand its mechanism for water oxidation and improve catalysis.

A recent homogeneous copper-based water oxidation catalyst, $[(66'bpy(OH)_2)Cu(OH)_2]^{2+}$, with a $66'bpy(OH)_2$ (6,6′-dihydroxy-2,2′-bipyridine) ligand (Figure 7.25) is hypothesized to mimic the role of tyrosine in Photosystem II.[58] Tyrosine in Photosystem II possesses a phenol group that forms a hydrogen bond with a nearby histidine imidazole, which enhances its oxidation ability through PCET.[7] A proton transfer to histidine coupled with an electron transfer to chlorophyll $P680^+$ avoids the formation of a high-energy tyrosine radical cation.[7] The hydroxyl groups of the ligand in the copper complex possess the ability to assist in the PCET processes, which aids in lowering overpotential and increasing the activity of the catalyst.[58] This complex remains significant to the development of future water oxidation catalysts based on copper because of large overpotentials associated with high oxidation state copper. The catalyst functions with a TOF of 0.4 s^{-1} and TON of 400.[58]

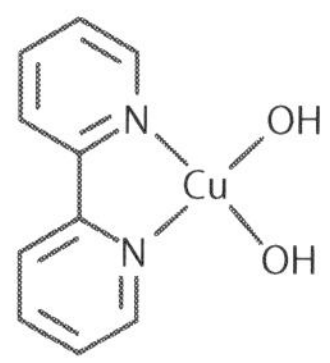

FIGURE 7.24 Structure of $[Cu(bpy)(OH)_2]^{2+}$ where bpy = 2,2′-bipyridine.

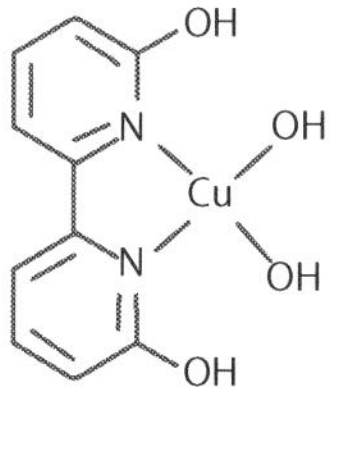

FIGURE 7.25 Structure of $[Cu(6,6'bpy(OH)_2)(OH)_2]^{2+}$ where $6,6'bpy(OH)_2$ = 6,6′-dihydroxy-bipyridine.

Therefore, this catalyst does not catalyze water oxidation as quickly as the copper-bipyridine-hydroxo complex, but it remains more stable and effective.

7.5.5 Iron

Iron remains one of the most abundant transition metals in nature, making its use in water oxidation catalysis less expensive than other alternatives.[19] In addition, iron possesses a number of different oxidation states and the ability to oxidize and reduce easily.[19] These qualities make the metal attractive for water oxidation catalysis because it must have the ability to access high oxidation states to bind water. In comparison to second- and third-row transition metals, such as ruthenium, platinum, and iridium, iron remains more biocompatible, cheaper, and safer.[61] Nature already uses iron to mediate proton/electron chemistry in hydrogenases. These enzymes consist of iron metal centers involved in electron transfer reactions and catalyze the reduction of hydrogen.[19] Therefore, iron possesses demonstrated qualities that make it attractive for use in water oxidation catalysis.

Iron has been incorporated into various homogeneous and heterogeneous catalysts that are capable of oxidizing water.[62–69] However, these catalysts require further mechanistic studies to enhance the stability and efficiency of the complexes. These catalysts possess lifetimes of several minutes, hours, or days and typically decompose to metal oxides during catalysis.[19] In addition to stability, the efficiency of iron-based heterogeneous catalysts requires improvement. Iron oxides possess less catalytic activity than cobalt oxides for water oxidation.[20] In some cases, the activity of the catalyst improves with the incorporation of other metals into the complex.[20]

An iron-based catalyst with a tetraamido macrocyclic ligand (Fe^{III}-TAML; Figure 7.26) activates oxygen as well as peroxides, providing evidence that these types of complexes possess the ability to take part in electron transfer reactions.[66] Because oxygen activation exists as the opposite of water oxidation, the principle of microscopic reversibility suggests that the forward and reverse reactions might possess similar pathways. Under certain conditions, these Fe-TAMLs complexes form Fe^{IV}-oxo and Fe^{V}-oxo compounds capable of water oxidation.[66] These complexes possess the capacity for incorporation into solar energy conversion devices due to low molecular weight and use of iron, which reduces cost.[66] In the presence of a

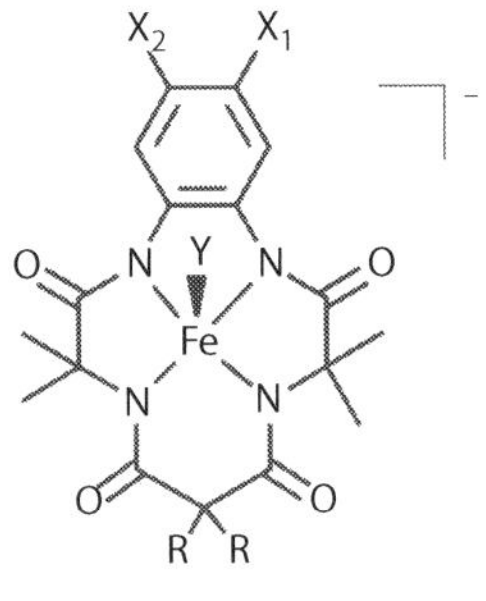

FIGURE 7.26 Structure of $[Fe^{III}\text{-TAMLs}]^-$ where TAML = tetraamido macrocyclic ligand and X_1/X_2 = H, NO_2, or Cl; R = CH_3, $(CH_2)_2$, or F; Y = H_2O.

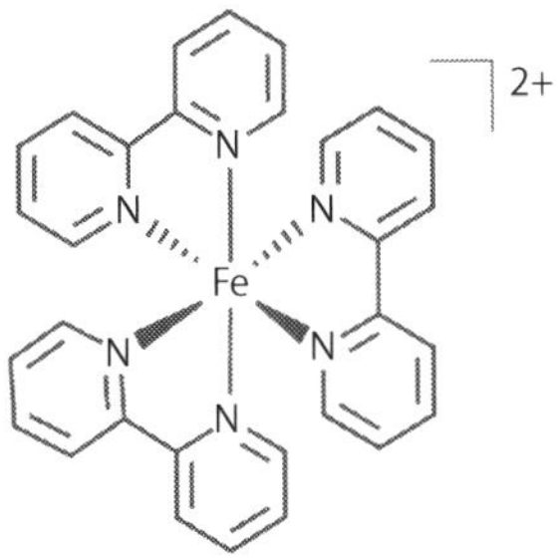

FIGURE 7.27 Structure of $[Fe(bpy)_3]^{2+}$ where bpy = 2,2′-bipyridine.

chemical oxidant, these catalysts possess a TON of 1.3 s^{-1}.[66] In the presence of a sacrificial oxidant, the catalyst maintains its activity for several hours.[66] Future work on Fe-TAMLs involves the development of a complex with the ability to perform light-driven water oxidation.

Another class of iron-based catalysts involves complexes with various ligands, including 2,2′-bipyridine (Figure 7.27), 2,2′:6′,2″-terpyridine (Figure 7.28), and ClO_4^- that form from iron salts in borate buffer.[62] $[Fe(bpy)_2Cl_2]Cl$, $[Fe(tpy)_2]Cl_2$, and $Fe(ClO_4)_3$ possess higher TONs (95, 19, 147, respectively) than the Fe-TAML complex (18).[62,66] These catalysts possess the ability to perform water oxidation in the presence of a chemical oxidant as well as visible light-driven water oxidation. The catalysts possess higher TONs under photochemical conditions (157, 376, 436, respectively), a significant benefit for solar energy conversion.[62]

During chemical and photochemical catalysis with all of these iron complexes, the formation of particles was detected. These particles were determined to be Fe_2O_3, so these nanoparticles were prepared and tested for catalytic activity.[62] These particles possess a TON of 58 for photochemical water oxidation and 152 for chemical water oxidation, which suggests that Fe^{III} converts into Fe_2O_3 nanoparticles under oxidative conditions and can participate in the water oxidation process.[62] These TONs remain similar to those of $Fe(ClO_4)_3$ under the same conditions, which provides evidence that the homogeneous iron catalysts may decompose, forming nanoparticles as a by-product that serve as the actual catalyst. Many molecular complexes function as active catalysts, but they may also serve as procatalysts to metal oxides that actually catalyze water oxidation.[2] In the latter case, the complex decomposes during catalysis, leaving particles behind that actually account for the catalytic activity.

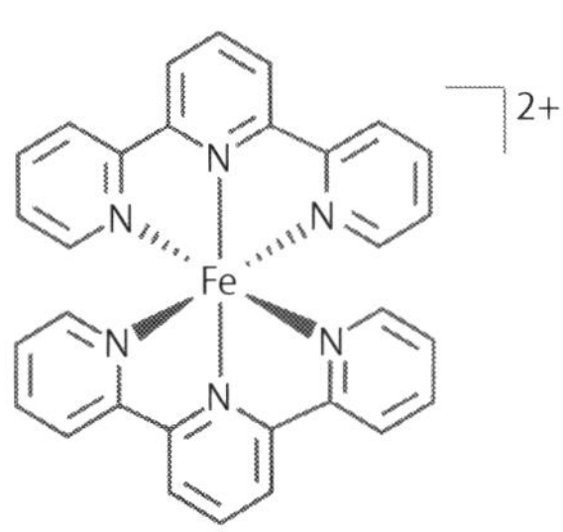

FIGURE 7.28 Structure of $[Fe(tpy)_2]^{2+}$ where tpy = 2,2′:6′2″-terpyridine.

7.5.6 Manganese

The OEC in Photosystem II contains a manganese cluster responsible for the water oxidation process. Therefore, manganese exists as an obvious choice for an economic water oxidation catalyst to use in an artificial photosynthetic system. Manganese complexes that possess comparable structures to or characteristics of the OEC may function similarly or provide information on its mechanism. In addition to iron, manganese exists as a highly earth-abundant transition metal.[52] Manganese also possesses a wide range of oxidation states, ranging from Mn^{II} to Mn^{VII}, which most likely accounts for the OEC's high efficiency.[70] The OEC possesses a TON of 180,000 and a TOF of 100 to 400 s^{-1}; therefore, the incorporation of manganese into artificial photosynthetic devices could possibly lead to highly productive systems.[18] The electrochemical potentials for these various oxidation states of manganese lie close to one another unlike other transition metals. Therefore, the transition to higher oxidation states occurs more easily, which leads to its high efficiency and catalytic activity.[70] Manganese exists as a better option for water oxidation catalysts compared with other transition metals because its overpotentials are generally lower.

Many manganese complexes have been synthesized in the hopes of creating a functional model of the OEC, but few possess the ability to oxidize water.[71] Homogeneous and heterogeneous catalysts based on manganese have been developed, but show less catalytic activity than complexes that use second-row transition metals.[18] The development of manganese oxides as water oxidation catalysts remains desirable to making a solar energy conversion device because of similarities to the metal-oxo core of the OEC; however, the catalytic activity of manganese oxides remains less than that of iridium oxides.[20] Still, the development of manganese versus iridium oxides remains more economically viable because of the former's earth abundance. In addition, the performance of manganese-based catalysts improves when doped onto a substrate or with another metal.[20] Therefore, continued research on manganese-based water oxidation catalysts remains significant to both understanding photosynthesis and the OEC as well as developing an economically viable catalyst with similar efficiency.

Homogeneous water oxidation catalysts that incorporate manganese metal have been developed, such as $[Mn_2^{III/IV}O_2(tpy)_2(H_2O)_2]^{3+}$ (Figure 7.29), which generates oxygen from water in aqueous solution.[72] This complex requires the use of a two-electron sacrificial oxidant to reach high valent manganese. Therefore, this complex cannot perform light-driven catalysis, which requires multiple one-electron steps.[72]

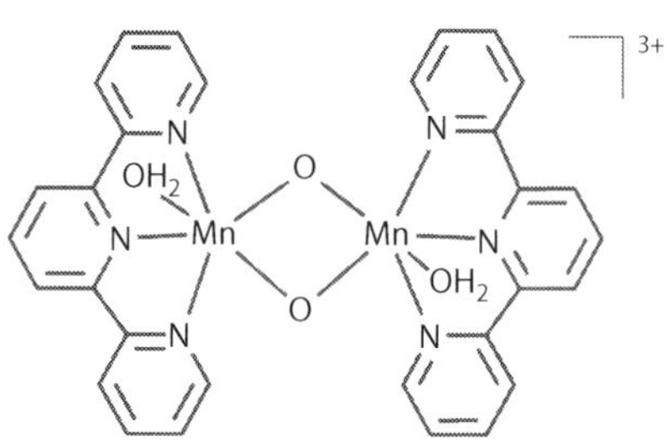

FIGURE 7.29 Structure of $\left[Mn_2^{III/IV}O_2(tpy)_2(H_2O)_2\right]^{3+}$ where tpy = 2,2′:6′2″-terpyridine.

Interestingly, immobilization of the catalyst in clay allows the complex to produce oxygen with a single electron oxidant.[72] This complex demonstrates that catalysts may activate in more than one way and function more effectively under certain conditions.

Two manganese tetramers, $[Mn_4^{IV}O_5(tpy)_4(H_2O)_2]^{6+}$ and $[Mn_4O_6(tacn)_4]^{4+}$ (tacn = 1,4,7-triazacyclononane), and manganese dioxide (MnO_2) deposit on electrodes and serve as water oxidation catalysts.[72] To increase the activity and stability of the manganese catalysts, the complexes are doped into a Nafion polymer. Nafion polymers possess acidic sulfonic acid head groups that become negatively charged and forms clusters capable of accepting positively charged complexes.[72] In addition, the Nafion functions as an immobilizing matrix, which "holds the manganese complexes near the electrode surface, stabilizes any reactive intermediates, and aids in proton management," which is similar to the D_1 protein in Photosystem II, stabilizing the transfer of protons and electrons.[72] These two complexes alone do not function as homogeneous catalysts, but with the support of the Nafion membrane, they serve as efficient heterogeneous catalysts.

A functional model of the manganese cluster in Photosystem II involves manganese oxide nanoparticles, which simulate the OEC and act as efficient water oxidation catalysts. The manganese calcium metal center in Photosystem II has a dimension of 0.5 nm, so the development of manganese nanoparticles would directly correlate to the structure of the OEC.[71] In the presence of a chemical oxidant, nanosized manganese oxides function as heterogeneous water oxidation catalysts with active sites on the surface.[71] The core of the OEC possesses a similar structure to that of a manganese oxide nanoparticle.

Manganese–calcium oxides exist as one of the best manganese-based water oxidation catalysts, truly imitating the manganese cluster involved in natural photosynthesis. The incorporation of calcium ions to manganese oxides improves their catalytic activity and provides an exemplary model of the OEC, which contributes to the current understanding of the natural process of water splitting.[71] Both calcium and manganese are inexpensive and earth-abundant elements, which makes manganese–calcium oxides a cost-efficient, functional model of the OEC with promise for artificial photosynthetic systems. The enhanced efficiency of these manganese–calcium oxides correlates to its "layered, open structure and small particle size."[71] Overall, manganese oxides possess attractive qualities, such as low cost, stability, and efficiency, which make the catalysts worthy candidates for artificial photosynthesis.

7.5.7 Nickel

In addition to the other first-row transition metals, nickel has been explored for its use in water oxidation catalysis. A previous use of nickel involves nickel oxides in anodic material for the generation of oxygen.[73] These nickel oxides have been used in large-scale water electrolyzers, serving as active and stable catalysts.[74] However, these electrolyzers function with undesirable conditions, such as a highly basic atmosphere.[19] With this background, nickel holds the potential for incorporation into a solar energy–generating device. Additional characteristics of nickel include "corrosion resistance, high catalytic activity, and earth abundance."[74] Nickel ions

(Ni^{II}) have also been used as dopants for other catalysts, such as manganese and iron oxides, which has improved their catalytic activity.[20] The Ni^{II} ions generate electron-rich d^8 systems, which possibly increase the oxidation power of other metals involved in the complex.[75]

A pentanickel silicotungstate complex, $K_{10}H_2[Ni_5(OH)_6(OH_2)_3(Si_2W_{18}O_{66})]\cdot 34H_2O$, catalyzes both chemical and light-driven water oxidation. The nickel-based POM consists of a metal–oxygen anion structure that stabilizes the nickel–oxygen core.[76] Both the nickel POM and the OEC possess a metal–oxygen structure consisting of earth-abundant materials. In pH 8.0 sodium borate buffer, the complex catalyzes light-driven water oxidation with a TON and TOF of 60 and approximately 1 s^{-1}, respectively.[76] The nickel POM remains molecular and does not decompose to nickel hydroxide particles.[76]

Similar to the Co-P_i catalyst, a nickel-based complex forms as a thin film from a solution of Ni^{II} ions and borate.[77] The nickel–borate (Ni-B_i) catalyst serves as an alternative to the Co-P_i catalyst, both of which utilize inexpensive materials, and operate under favorable conditions. The Ni-B_i catalyst differs from Co-P_i because it requires a second oxidation step to achieve catalytic activity.[78] Ni^{II} ions deposit as Ni^{III}, an inactive form of the catalyst, which gets coated through electrolysis to form active Ni^{IV} ions.[78] Both of these complexes serve as heterogeneous catalysts with potential for solar energy conversion. The nickel films possess the ability for thickness adjustment and function with high catalytic activity.[78]

7.5.8 Mixed Metal Systems

The previous sections of this chapter have discussed single metal center complexes that act as water oxidation catalysts. A number of current research efforts explore the effects of incorporating multiple metal atoms into one crystal structure in an attempt to lower the energy of their conduction band while also increasing their catalytic activity toward water splitting.[79,80] The Mn_3O_4Ca cluster system serves as an example of this effect, in which the manganese and calcium ions were insufficient catalysts on their own, but together the performance of the system significantly increased.[8] Therefore, current research efforts involving mixed metal systems focus on the investigation of perovskite metal oxide systems as potential water oxidation catalysts.[79,80]

Perovskite metal oxide systems adopt the chemical formula of ABO_3, in which "A" typically represents an alkaline earth metal and "B" typically represents a transition metal, although rare earth metals serve as potential alternatives for the B site of these materials.[81,82] Perovskites possess the ability to accommodate oxygen vacancies and limited stoichiometric variation in the crystal structure, while maintaining structural stability.[81,82] These abilities result in tunable perovskite properties, which allows for optimization of catalytic activity by inducing either *n*-type (doped with an electron donor) or *p*-type (doped with an electron acceptor) conductivity with the addition of various dopants into the A and B sites of the crystal system, that is, $A'_{1-x}A''_xB'_{1-y}B''_yO_{3-\delta}$.[81,82] The ability to form oxygen ion–specific vacancies in the crystal structure remains imperative to transport the oxide anions once they are generated from the splitting of water. These complexes must also exhibit suitable electron conductivity to transport electrons in the external circuit from the oxidation of oxide anions for use in the reduction of hydrogen cations to form H_2.

Recent perovskite metal oxide systems use the manganese–calcium–based clusters that occur in nature for photosynthesis as a model. Patra et al.[80] have investigated the effects of doping strontium into lanthanum manganites to form $(La_{1-x}Sr_x)MnO_{3-\delta}$ $(0.0 \leq x \leq 0.5)$. They observe that with an increase in strontium dopants, they could form oxygen vacancies by substituting the lanthanum with an oxidation state of 3+ and strontium with an oxidation state of 2+.[80] The increase in ionic radius of strontium to lanthanum also caused a distortion in the crystal lattice with an increase in Sr dopant concentration that allowed for the Mn-O-Mn bond angle to reach an optimum value of 180° for maximum charge transport through the system.[80] Some research efforts also extend past the typical manganese–calcium systems. For example, Suntivich et al.[79] found that a cathode material for solid oxide fuel cells, $Ba_{0.5}Sr_{0.5}Co_{0.8}Fe_{0.2}O_{3-\delta}$, outperforms other manganese- and calcium-based systems as water oxidation catalysts.

7.6 CONCLUSIONS

Discovering systems based on transition metals that can efficiently and robustly catalyze the oxidation of water to molecular oxygen would open the door to potential renewable energy sources. Furthermore, if these catalysts can utilize sunlight to provide the energy for this reaction, a truly sustainable means of securing our energy future remains attainable. Although many advances have been made in the field of water oxidation, this area of study remains highly active with many avenues yet to explore. For example, the exploration of first-row transition has led to numerous complexes capable of water oxidation catalysis. Even though these metals remain desirable for practical applications due to low cost, other qualities (i.e., stability, activity, efficiency, and robustness) require improvement. Today, research efforts focused on the enhancement of these catalysts remain prevalent in the field of water oxidation catalysis. A better understanding of the water oxidation mechanism in Photosystem II may aid in the improvement of these catalysts for artificial photosynthetic devices. In addition, coupling the process of light-driven water oxidation with catalysts that can utilize the reductive equivalents to form storable and clean fuel sources continues to be a major focal point moving forward.

REFERENCES

1. J. Rifkin. *The Hydrogen Economy*, Penguin Group Inc., New York, 2002, pp. 37–64.
2. H. Liv, Y. V. Geletii, C. Zhao, J. W. Ickers, G. Zhu, Z. Luo, J. Song, T. Lian, D. G. Musaev, and C. L. Hill. *Chem. Soc. Rev.* 41 (2012) 7572–7589.
3. N. Lenssen, and C. Flavin. *Energ Policy* 24 (1996) 769–781.
4. T. T. Chow, G. Pei, K. F. Fong, Z. Lin, A. L. S. Chan, and J. Ji. *Appl Energ* 86 (2009) 310–316.
5. C. J. Gagliardi, B. C. Westlake, C. A. Kent, J. J. Paul, J. M. Papanikolas, and T. J. Meyer. *Coord. Chem. Rev.* 254 (2010) 2459–2471.
6. J. Barber. *Chem. Soc. Rev.* 38 (2009) 185–196.

7. D. R. Weinberg, C. J. Gagliardi, J. F. Hull, C. F. Murphy, C. A. Kent, B. C. Westlake, A. Paul, D. H. Ess, D. G. McCafferty, and T. J. Meyer. *Chem. Rev.* 112 (2012) 4016–4093.
8. Y. Umena, K. Kawakami, J.-R. Shen, and N. Kamiy. *Nature* 473 (2011) 55–60.
9. J. P. McEvoy, and G. W. Brudvig. *Chem. Rev.* 106 (2006) 4455–4483.
10. J. S. Kanady, E. Y. Tsui, M. W. Day, and T. Agapie. *Science* 333 (2011) 733–736.
11. E. Y. Tsui, J. S. Kanady, and T. Agapie. *Inorg. Chem.* 52 (2013) 13833–13848.
12. A. W. Rutherford, J. E. Mullet, and A. R. Crofts. *FEBS Lett.* 123 (1981) 235–237.
13. W. Ruttinger, and G. C. Dismukes. *Chem. Rev.* 97 (1997) 1–24.
14. A. Dovletoglou, S. A. Adeyemi, and T. J. Meyer. *Inorg. Chem.* 35 (1996) 4120–4127.
15. J. R. Swierk, and T. E. Mallouk. *Chem. Soc. Rev.* 42 (2013) 2357–2387.
16. C. W. Cady, R. H. Crabtree, and G. W. Brudvig. *Coord. Chem. Rev.* 252 (2008) 444–455.
17. H. Yamazaki, A. Shouji, M. Kajita, and M. Yagi. *Coord. Chem. Rev.* 254 (2010) 2483–2491.
18. A. Singh, and L. Spiccia. *Coord. Chem. Rev.* 257 (2013) 2607–2622.
19. P. Du, and R. Eisenberg. *Energy Environ. Sci.* 5 (2012) 6012–6021.
20. S. Fukuzumi, D. Hong, and Y. Yamada. *J. Phys. Chem. Lett.* 4 (2013) 3458–3467.
21. L. L. Tinker, N. D. McDaniel, and S. Bernhard. *J. Mater. Chem.* 19 (2009) 3328–3337.
22. S. Campagna, F. Puntoriero, F. Nastasi, G. Bergamini, and V. Balzani. *Top. Curr. Chem.* 280 (2007) 117–214.
23. M. Yagi, and M. Kaneko. *Chem. Rev.* 101 (2001) 21–35.
24. Q. Yin, J. M. Tan, C. Besson, Y. V. Geletii, D. G. Musaev, A. E. Kuznetsov, Z. Luo, K. I. Hardcastle, and C. L. Hill. *Science* 328 (2010) 342–345.
25. S. Fukuzumi, and D. Hong. *Eur. J. Inorg. Chem.* (2014) 645–659.
26. J. J. Concepcion, J. W. Jurss, M. K. Brennaman, P. G. Hoertz, A. O. T. Patrocinio, N. Y. M. Iha, J. L. Templeton, and T. J. Meyer. *Acc. Chem. Res.* 42 (2009) 1954–1965.
27. J. A. Gilbert, D. S. Eggleston, W. R. Murphy, D. A. Geselowitz, S. W. Gersten, D. J. Hodgson, and T. J. Meyer. *J. Am. Chem. Soc.* 107 (1985) 3855–3864.
28. F. Liu, J. J. Concepcion, J. W. Jurss, T. Cardolaccia, J. L. Templeton, and T. J. Meyer. *Inorg. Chem.* 47 (2008) 1727–1752.
29. J. P. Collin, and J. P. Sauvage. *Inorg. Chem.* 25 (1986) 135–141.
30. K. Nagoshi, S. Yamashita, M. Yagi, and M. Kaneko. *J. Mol. Catal. A: Chem.* 144 (1999) 71–76.
31. R. Bianco, P. J. Hay, and J. T. Hynes. *J. Phys. Chem. A.* 115 (2011) 8003–8016.
32. D. Moonshiram, V. Purohit, J. J. Concepcion, T. J. Meyer, and Y. Pushkar. *Materials* 6 (2013) 392–409.
33. J. J. Concepcion, J. W. Jurss, J. L. Templeton, and T. J. Meyer. *J. Am. Chem. Soc.* 130 (2008) 16462–16463.
34. R. Zong, and R. P. Thummel. *J. Am. Chem. Soc.* 127 (2005) 12082–12083.
35. D. C. Marelius, S. Bhagan, D. J. Charboneau, K. M. Schroeder, J. M. Kamadar, A. R. McGettigan, B. J. Freeman et al. *Eur. J. Inorg. Chem.* 2014 (4) 676–689.
36. Z. Chen, J. J. Concepcion, J. W. Jurss, and T. J. Meyer. *J. Am. Chem. Soc.* 131 (2009) 15580–15581.
37. H.-W. Tseng, R. Zong, J. T. Muckerman, and R. Thummel. *Inorg. Chem.* 47 (2008) 11763–11773.
38. N. Kaveevivitchai, R. Zong, H.-W. Tseng, R. Chitta, and R. P. Thummel. *Inorg. Chem.* 51 (2012) 2930–2939.
39. R. Cao, W. Lai, and P. Du. *Energy Environ. Sci.* 5 (2012) 8134–8157.
40. J. J. Concepcion, J. W. Jurss, M. R. Norris, Z. Chen, J. L. Templeton, and T. J. Meyer. *Inorg. Chem.* 49 (2010) 1277–1279.
41. N. D. McDaniel, F. J. Coughlin, L. L. Tinker, and S. Bernhard. *J. Am. Chem. Soc.* 130 (2008) 210–217.

42. A. Petronilho, M. Rahman, J. A. Woods, H. Al-Sayyed, H. Muller-Bunz, J. M. D. MacElroy, S. Bernhard, and M. Albrecht. *Dalton Trans.* 41 (2012) 13074–13080.
43. J. F. Hull, D. Balcells, J. D. Blakemore, C. D. Incarvito, O. Eisenstein, G. W. Brudvig, and R. H. Crabtree. *J. Am. Chem. Soc.* 131 (2009) 8730–8731.
44. J. D. Blakemore, N. D. Schley, D. Balcells, J. F. Hull, G. W. Olack, C. D. Incarvito, O. Eisenstein, G. W. Brudvig, and R. H. Crabtree. *J. Am. Chem. Soc.* 132 (2010) 16017–16029.
45. R. H. Crabtree. *Science* 330 (2010) 455–456.
46. K. A. Maghacut, A. B. Wood, W. J. Boyko, T. J. Dudley, and J. J. Paul. *Polyhedron* 67 (2014) 329–337.
47. K. T. Hufziger, F. S. Thowfeik, D. J. Charboneau, I. Nieto, W. G. Dougherty, W. S. Kassel, T. J. Dudley, E. J. Merino, E. T. Papish, and J. J. Paul. *J. Inorg. Biochem.* 130 (2014) 103–111.
48. M. J. Fuentes, R. J. Bognanno, W. G. Dougherty, W. J. Boyko, W. S. Kassel, T. J. Dudley, and J. J. Paul. *Dalton Trans.* 41 (2012) 12514–12523.
49. S. Klein, W. G. Dougherty, W. S. Kassel, T. J. Dudley, and J. J. Paul. *Inorg. Chem.* 50 (2011) 2754–2763.
50. J. DePasquale, I. Nieto, L. E. Reuther, C. J. Herbst-Gervasoni, J. J. Paul, V. Mochalin, M. Zeller, C. M. Thomas, A. W. Addison, and E. T. Papish. *Inorg. Chem.* 52 (2013) 9175–9183.
51. A. Lewandowska-Andralojc, D. E. Polyansky, C.-H. Wang, W.-H. Wang, Y. Himeda, and E. Fujita. *Phys. Chem. Chem. Phys.* 16 (2014) 11976–11987.
52. J. J. R. Frausto da Silva, and R. J. P. Williams. *The Biological Chemistry of the Elements: The Inorganic Chemistry of Life.* Oxford University Press, New York, 2001, pp. 341–397, 418–432, 436–448.
53. J. G. McAlpin, T. A. Stich, W. H. Casey, and R. D. Britt. *Coord. Chem. Rev.* 256 (2012) 2445–2452.
54. V. Artero, M. Chavarot-Kerlidou, and M. Fontecave. *Angew. Chem. Int. Ed.* 50 (2011) 7238–7266.
55. M. W. Kanan, Y. Surendranath, and D. G. Nocera. *Chem. Soc. Rev.* 38 (2009) 109–114.
56. M. W. Kanan, and D. G. Nocera. *Science* 321 (2008) 1072–1075.
57. E. A. Lewis, and W. B. Tolman. *Chem. Rev.* 104 (2004) 1047–1076.
58. T. Zhang, C. Wang, S. Liu, J.-L. Wang, and W. Lin. *J. Am. Chem. Soc.* 136 (2013) 273–281.
59. Z. Chen, and T. J. Meyer. *Angew. Chem. Int. Ed.* 52 (2013) 700–703.
60. S. M. Barnett, K. I. Goldberg, and J. M. Mayer. *Nat. Chem.* 4 (2012) 498–502.
61. T. W.-S. Chow, G.-Q. Chen, Y. Liu, C.-Y. Zhou, and C.-M. Che. *Pure Appl. Chem.* 84 (2012) 1685–1704.
62. G. Chen, L. Chen, S.-M. Ng, W.-L. Man, and T.-C. Lau. *Angew. Chem. Int. Ed.* 52 (2013) 1789–1791.
63. A. Poater, *Catal. Commun.* 44 (2014) 2–5.
64. M. M. Najafpour, A. N. Moghaddam, D. J. Sedigh, and M. Hotynska. *Catal. Sci. Technol.* 4 (2014) 30–33.
65. D. Hong, S. Mandal, Y. Yamada, Y.-M. Lee, W. Nam, A. Llobet, and S. Fukuzumi. *Inorg. Chem.* 52 (2013) 9522–9531.
66. W. C. Ellis, N. D. McDaniel, S. Bernhard, and T. J. Collins. *J. Am. Chem. Soc.* 132 (2010) 10990–10991.
67. D. Hong, Y. Yamada, T. Nagatomi, Y. Takai, and S. Fukuzumi. *J. Am. Chem. Soc.* 134 (2012) 19572–19575.
68. M. Gong, Y. Li, H. Wang, Y. Liang, J. Z. Wu, J. Zhou, J. Wang, T. Regier, F. Wei, and H. Dai. *J. Am. Chem. Soc.* 135 (2013) 8452–8455.

69. J. L. Fillol, Z. Codola, I. Garcia-Bosch, L. Gomez, J. J. Pla, and M. Costas. *Nat. Chem.* 3 (2011) 807–813.
70. G. L. Elizarova, G. M. Zhidomirov, and V. N. Parmon. *Catal. Today* 58 (2000) 71–88.
71. M. M. Najafpour, F. Rahimi, E.-M. Aro, C.-H. Lee, and S. I. Allakhverdiev. *J. R. Soc. Interface* 9 (2012) 2383–2395.
72. K. J. Young, Y. Gao, and G. W. Brudvig. *Aust. J. Chem.* 64 (2011) 1221–1228.
73. G. Chen, L. Chen, S.-M. Ng, and T.-C. Lau. *Chem. Sus. Chem.* 7 (2014) 127–134.
74. M. J. Kenney, M. Gong, Y. Li, J. Z. Wu, J. Feng, M. Lanza, and H. Dai. *Science* 342 (2013) 836–840.
75. D. Hong, Y. Yamada, A. Nomura, and S. Fukuzumi. *Phys. Chem. Chem. Phys.* 15 (2013) 19125–19128.
76. G. Zhu, E. N. Glass, C. Zhao, H. Lv, J. W. Vickers, Y. V. Geletii, D. G. Musaev, J. Song, and C. L. Hill. *Dalton Trans.* 41 (2012) 13043–13049.
77. M. Dinca, Y. Surendranath, and D. G. Nocera. *Proc. Natl. Acad. Sci. U.S.A.* 107 (2010) 10337–10341.
78. D. K. Bediako, B. Lassalle-Kaiser, Y. Surendranath, J. Yano, V. K. Yachandra, and D. G. Nocera. *J. Am. Chem. Soc.* 134 (2012) 6801–6809.
79. J. Suntivich, K. May, H. Gasteiger, J. Goodenough, and Y. Saho-Horn. *Science* 334 (2011) 1383–1385.
80. A. Patra, N. Kumar, D. Barpuzary, M. De, and M. Qureshi. *Mater. Lett.* 131 (2014) 124–127.
81. M. Gong, X. Liu, and J. Trembly. *J. Power Sources* 168 (2007) 289–298.
82. R. Huang, Z. Dass, J. Xing, and J. Goodenough. *Science* 312 (2006) 254–257.

8 Life Cycle Thinking Informs Catalysis Choice and Green Chemistry

Philip Nuss

CONTENTS

Let us not demand more of the Earth.
Let us do more with what the Earth provides.

Gunter Pauli

8.1 INTRODUCTION

Since the late eighteenth century, humans have been altering the earth at an unprecedented and unsustainable rate and scale.[1] Our economies are based on global resource use with modern man consuming between 30 and 75 tons of material per person per year in their companies and households.[2] Of the materials consumed, an estimated 90% of all biomass inputs and more than 90% of the nonrenewable materials used are wasted on the way to providing products to the end-user.[3] Resource extraction can result in serious environmental damages through the extraction and refining processes itself, and also due to the increasing transport distances between extraction, processing and final consumption. The chemical and petrochemical industry is a large user of chemical feedstock and energy (10% of total final energy demand) and contributes approximately 7% to global greenhouse gas (GHG) emissions.[4] However, chemical products and technologies are also used in a wide variety of applications (e.g., insulation or lighter and advanced materials for transport) that help conserve resources and energy, thereby reducing environmental pressures. With the goal to develop more cost-effective and environmentally benign processes, the chemicals industry is increasingly seeking to replace stoichiometric reagents with catalytic routes. Today, an estimated 60% of chemical products and 90% of chemical processes employ catalysts,[5] as do nearly all petroleum-refining processes.[6] Their role in chemical synthesis is to facilitate reaction pathways with lower activation energies and to avoid the production of unwanted by-products. Because catalysis allows for easier-to-control process conditions (e.g., operation at lower temperatures and pressures or increased yields), it finds use in a wide range of applications including the production of commodity and petrochemicals, fine chemicals, pharmaceuticals, and food products.[7,8] Catalysis can be divided into heterogeneous (acting in a different phase than the reactants) and homogeneous (acting in the same phase as the reactants) catalysts. The use of enzymes for chemical transformations is referred to as biocatalysis. In the future, the use for catalysis is expected to further increase, especially in environmental catalysis (e.g., for the reduction of sulfur and NO_x emissions), alternative fuel and energy systems (e.g., hydrogen cells and biofuels), and water purification and recycling.[9,10]

Because catalysis allows for more efficient, less energy-intensive, and more selective chemical reactions, it is regarded by many as an enabling technology to promote overall sustainability.[11–14] The concept of "sustainability" was defined in 1987 by the Brundtland Commission as "Sustainable development meets present needs without compromising the ability of future generations to meet their needs."[15]

Following this widely accepted definition of sustainable development, Christensen[16] defines sustainable chemistry as "chemistry that contributes to securing the needs of the present without compromising the ability of future generations to meet their own needs." Consequently, sustainable chemistry should consider the three dimensions of sustainability displayed schematically in Figure 8.1.[17]

Because chemistry is pervasive in so many aspects of our societal goals (e.g., improved materials, healthcare, agrochemicals, and energy), catalysis possesses the potential to reduce environmental pressures and enhance overall sustainability. Indeed, the International Energy Agency estimates that catalytic and related process improvements in the chemical industry could reduce energy consumption by 13 exajoules (EJ) (equivalent to the current annual primary energy use of Germany) and 1 gigatonne (Gt) of carbon dioxide equivalent (CO_2-eq) per year by 2050.[4] However, a systematic assessment taking into account various environmental pressures, and ideally the various aspects of sustainability, is required to avoid burden shifting and improve system-wide performance.

This book chapter first introduces a number of system analysis tools (e.g., material flow analysis [MFA] and life cycle assessment [LCA]) used to quantify material stocks and flows and associated environmental effects along a product's life cycle. Sustainable development requires not only the piecemeal improvements of separate parts of production, consumption, and waste management but must also make sure that, for example, environmental impacts or anthropogenic material flows, are reduced from a system-wide perspective. The term *system* or *life cycle*

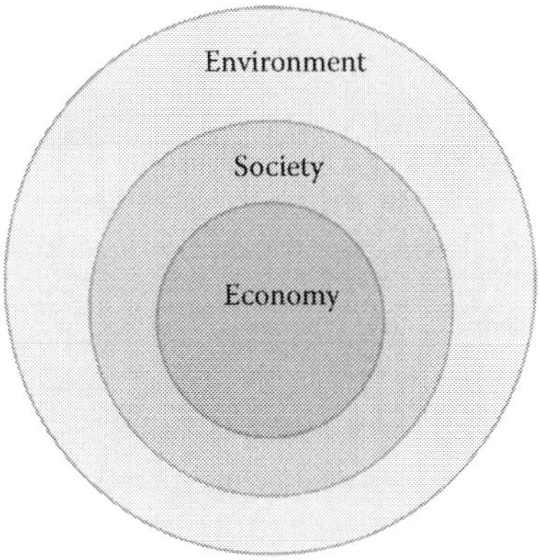

FIGURE 8.1 A diagram showing the three spheres of sustainability. Economy and society are constrained by environmental limits, and the economy is a construct of society. (Adapted from Mebratu, D., *Environ. Impact Assess. Rev.*, 18, 493–520, 1998.)

refers to the major activities across the life span of a material, product, process, or service beginning with resource extraction (mining, logging, and raw material extraction) to materials processing, to manufacturing and fabrication, to use, and then to collection, processing, and ultimate end-of-life management. Our focus is on environmental impacts, but a few references regarding social and economic approaches in the system analysis toolbox are also provided in the next section. We then continue to discuss the use of these tools in the context of sustainable (green) chemistry and catalysis. The chapter concludes with a case study applying LCA to bioacrylics.

8.2 LIFE CYCLE THINKING INFORMS GREENER PRODUCTS

Lack of availability or consideration of life cycle data in product design and decision-making has created unintended environmental consequences, some of which have only been recognized in recent years, such as indirect land use change in first-generation biofuels production.[18] Today, a number of tools are available to assess the life cycle–wide performance of a new process or product chain. These include, for example, MFA,[19] environmental LCA,[20,21] environmentally extended input/output (EIO) analysis,[22] life cycle cost (LCC) analysis,[23] social life cycle assessment (SLCA),[24] and resource criticality assessments.[25–27] Life cycle thinking reflects the considerations of cradle-to-grave implications of any action, and common to all these concepts is that they are taking a systems approach, considering upstream and downstream processes.[28]

8.2.1 Material Flow Analysis

MFA aims at quantifying the stocks and flows of materials at different temporal and spatial scales (e.g., an economy, country, region, community, business, company, or household[19,29] to study their role in the industrial metabolism*.[30] Materials can refer to bulk materials (e.g., steel, wood, and total mass), but also to single substances (e.g., a metal) or group of substances (e.g., the rare earth elements).† In the case of metals, the material cycles are expressed through four principal processes: production, fabrication and manufacturing, use, and waste management and recycling. A cycle is characterized by processes that are linked through markets,[31] each market indicating trade with other regions at the respective life stages. The scrap market plays a central role in that it connects manufacturing and waste management and recycling with production and fabrication. The cycle is surrounded by entities lying outside the system boundary: trade partners (other regions), Earth's crust from which ore extraction takes place, and repositories for metals in production waste deposits and landfills. Understanding the whole system of material flows can help quantify potential primary and secondary source strengths, manage resource use more wisely, and protect the environment.

In the context of catalysis, the platinum group metals (PGMs) consisting of ruthenium, rhodium, palladium, osmium, iridium, and platinum present an interesting case in which MFA elucidates potential burden shifting.‡ The PGMs are among the rarest elements in Earth's upper continental crust[32] and their production is resource-intensive and energy-intensive.[33] Major PGM end-use applications include autocatalysts (for the automobile industry), chemical processes (e.g., use in the chemical and petroleum industries), dentistry, jewelry, electronics, investment, the glass industry, and others.[34] Their role in automobile catalysts is to reduce emissions, for example, of nitrous oxides (NO_x), during the vehicle driving phase. However, because their production is resource-intensive, obtaining PGMs may also result in environmental pressures in the major mining countries of Russia, South Africa, and North America. Indeed, a MFA study by Saurat and Bringezu[35] showed that using PGM catalysts in European cars results in a significant overall reduction of acidifying emissions (SO_2-eq) when compared with cars without using a catalytic converter, whereas life cycle–wide effects to global warming potential (GWP expressed in CO_2-equivalents) are slightly lower for gasoline cars and significantly higher for diesel cars (Figure 8.2). The analysis also indicates that the use of car catalysts leads to a drastic increase in resource extraction and mining waste (expressed in total material requirement [TMR]; Figure 8.2).

* The term *metabolism*, applied to plants or animals, includes the transformations of inputs (sunlight, chemical energy, nutrients, water, air) required by an organism to function and associated waste products. "Industrial metabolism," by analogy, refers to the flows of materials, energy, and waste in industrial systems.

† A group of metals including lanthanum (La), cerium (Ce), praseodymium (Pr), neodymium (Nd), samarium (Sm), europium (Eu), gadolinium (Gd), terbium (Tb), dysprosium (Dy), holmium (Ho), erbium (Er), thulium (Tm), ytterbium (Yb), lutetium (Lu), and yttrium (Y) used increasingly in modern technology.

‡ In the context of this chapter, burden shifting refers to attempts to improve the environmental performance or material efficiencies in one part of a product system at the expense of increasing the effects elsewhere. Only if the full life cycle (from "cradle-to-grave") is taken into consideration can burden shifting be elucidated and addressed. Burden shifting can take place in various ways, for example, between different media, between geographical regions, between life cycle stages, between generation, or between different effect categories (see, e.g., Wrisberg and de Haes[28] for a summary).

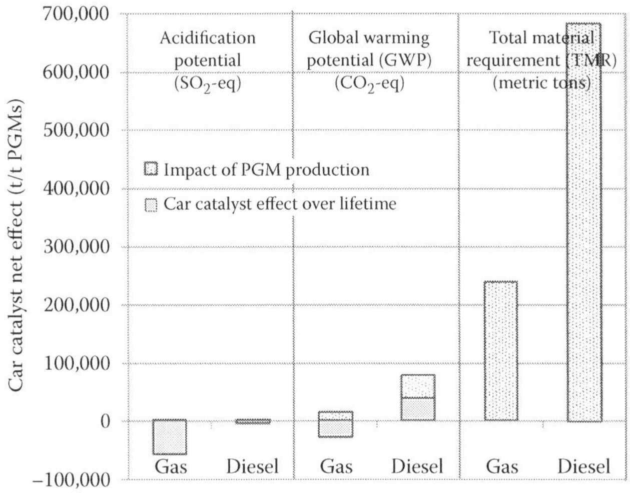

FIGURE 8.2 Net effect of gas or diesel car catalysts over their service lifetime (i.e., in comparison to cars without a catalytic converter). The negative effects are due to the emissions avoided from cars using a PGM catalyst when compared with cars without a catalytic converter. eq = equivalents. (Data from Saurat, M., Bringezu, S., *J. Ind. Ecol.*, 12, 754–767, 2008.)

Key point: Use of car catalysts in Europe can result in shifting emissions to other locations.

Although the intention of the catalytic converter is to reduce tailpipe emissions and associated human health and ecosystem damages, a more system-wide examination shows that the use of catalytic converters in automobiles in Europe simply shifts some of the environmental pressures to the PGM-producing countries.[35] In addition, the data indicate a shift from emissions-related effects (acidification potential and GWP) toward potential effects due to raw materials movement (TMR).[36] One possible strategy to enhance the positive effect of catalytic converters and avoid shifting environmental effects to other locations is to increase recycling rates for PGMs to replace primary by secondary production.[35] However, to date, the growing stock of cars cannot be matched by secondary production.

8.2.2 Life Cycle Assessment

LCA is a tool to systematically evaluate the potential cradle-to-grave environmental impacts of products, technologies, and services. The LCA concept can be traced back to the 1960s, when its forerunners (i.e., resource and environmental profile analysis and net energy analysis) were developed. Today, LCA as broadly defined by the International Organization for Standardization[20,21] consists of four steps: first, the goal and scope explains the study context and system boundaries, and defines a (quantifiable) functional unit to describe the primary function(s) fulfilled by the product or service system. Second, a life cycle inventory (LCI) comprising all environmentally relevant flows, such as material and energy inputs and outputs and emissions to air, soil, and water, is compiled. Third, the LCI is translated into potential effects including damage to human health, ecosystems, and depletion of resources, using life cycle impact assessment (LCIA) models. Finally, during the interpretation stage, results of the LCA and their wider implications are critically assessed. Recent efforts have been focused on broadening the traditional LCA framework to integrate environmental, social,[24] and economic[23,37] aspects at varying spatial levels, also referred to as life cycle sustainability assessment (LCSA).[38] A concise overview of LCA is given, for example, in Hellweg and Canals[39] and a more detailed explanation provided elsewhere.[40,41]

In the context of catalysis, LCA has been applied to compare environmental impacts of different catalytic reaction processes.[42–44] For example, Holman et al.[42] investigated the environmental impacts of acrylic acid (2-propenoic acid) production from propylene in a two-step reaction via acrolein with the use of propane as a potential starting material in a one-step process through a selective oxidation reaction (see the following equation).

$$CH_2 = CHCH_3 \text{ (propylene)} + O_2 \rightarrow CH_2 = CHCHO \text{ (acrolein)} + H_2O$$

$$CH_2 = CHCHO + 1/2\ O_2 \rightarrow CH_2 = CHCO_2H \text{ (acrylic acid)}$$

Acrylic acid is a commodity chemical with a market that reached approximately 3.2 million tons by the end of 2005.[45] Acrylic acid finds application, for example, in superabsorbent materials (e.g., for diapers) and as a replacement for phosphates in detergents. However, the propane process has not been commercialized and current work focuses on the development of catalysts to increase the conversion of propane and the selectivity toward acrylic acid. The use of LCA showed that even at low yields exceeding 33%, the propane process may have a lower GWP than the current propylene process (compared with a yield of 87% of the current commercial propylene process). This is mostly due to the upstream environmental burdens associated with propylene production, which are higher than those of propane production. Thus, from an environmental standpoint, the propane process may already present a viable option even at low conversion rates using current catalysts.

Key point: LCA can help to illustrate the environmental impacts of alternative catalyst-based processing routes, even when one of these routes has not yet been fully developed.

8.2.3 Metal Catalysis and Environmental Impacts

Organometallic catalysts are important in both fine and bulk chemicals production. However, most of the catalysts employed today rely on rare and expensive transition metals with potentially high toxicity and life cycle–wide environmental impacts.[11,46] Furthermore, because these metals cannot be metabolized by biological systems, any resulting catalyst waste may cause environmental problems in the long-term.[11]

As a result, researchers are increasingly looking into alternatives including (a) biocatalysis, (b) organocatalysis, and (c) sustainable metal catalysis. The term "sustainable metal catalysis" refers to the use of metals that are nontoxic, abundant, and biologically relevant, for example, magnesium, calcium, manganese, iron, cobalt, nickel, copper, zinc, and molybdenum.[11] An example is metal-catalyzed olefin hydrosilation, in which silicon–hydrogen bonds are added across a carbon–carbon double bond—a process used in the commercial production of silicone-based surfactants, fluids, molding products, coatings, and additives.[47] For many years, platinum-based compounds such as Karstedt's and Speier's catalysts were used by industry, whereas iron catalysts were introduced only recently as a potential alternative.[48] Indeed, a review of metal LCAs showed that on a per kilogram basis, iron production results in a much lower GWP (1.5 kg CO_2-eq/kg) when compared with platinum production (~12,500 kg CO_2-eq/kg; Figure 8.3).[49]

Key point: LCA can provide interesting insights into the environmental implications of the choice of catalytic metals already in the design phase.

Using such LCA-based information can guide metal catalyst developers in choosing a metal with the lowest environmental impact when designing new metal catalytic compounds. However, only one effect category is shown in Figure 8.3, and ideally, the catalysts' designer should also look at additional effect categories such as toxicity, cumulative energy demand, water use, acidification, and eutrophication, and carry out a cost analysis[50] when making the choice of using one metal over another. In some cases, the ligands surrounding the metal in homogeneous catalysis are expensive[51] and might also contribute to environmental impacts, for example, if a very resource-intensive chemical is used to produce them. A clear limitation of a cradle-to-gate LCA, as shown in Figure 8.3, is that the use and end-of-life phases are not included within the system boundary. For example, a catalyst may have a high environmental impact in the production phase but display only minor loss rates during the use phase and high recycling rates at end-of-life. This would only be captured by doing a cradle-to-grave assessment. For example, it is estimated that in 2007, the silicone industry consumed approximately 5.6 metric tons of platinum catalyst with most of this not being recovered.[52] Taking the result from Figure 8.3, this equals 70,000 metric tons CO_2-eq in 2007 (the per kilogram effect reported for platinum in Figure 8.3 already includes secondary production of the global supply mix). This figure could be drastically reduced if platinum catalysts were recycled at end-of-life. Nevertheless, if used together with cost and detailed energy analysis,[50] detailed information on the cradle-to-gate environmental impacts can provide a first indication of environmental impacts already in the design phase of a new metal catalyst as well as how an alternative option might compare with the metals currently used.

8.2.4 Criticality Assessment

However, LCA only accounts for environmental impacts and does not yet cover other important issues such as social and regulatory and geopolitical factors of

Global warming potential (kg CO_2-eq/kg)

Lowest — Highest

H																	He 0.9
Li 7.1	Be 122											B 1.5	C	N	O	F	Ne
Na	Mg 5.4											Al 8.2	Si	P	S	Cl	Ar
K	Ca 1.0	Sc 5710	Ti 8.1	V 33.1	Cr 2.4	Mn 1.0	Fe 1.5	Co 8.3	Ni 6.5	Cu 2.8	Zn 3.1	Ga 205	Ge 170	As 0.3	Se 3.6	Br	Kr
Rb	Sr 3.2	Y 15.1	Zr 1.1	Nb 12.5	Mo 5.7	Tc	Ru 2110	Rh 35,100	Pd 3880	Ag 196	Cd 3.0	In 102	Sn 17.1	Sb 12.9	Te 21.9	I	Xe
Cs	Ba 0.2	La-Lu*	Hf 131	Ta 260	W 12.6	Re 450	Os 4560	Ir 8860	Pt 12,500	Au 12,500	Hg 12.1	Tl 376	Pb 1.3	Bi 58.9	Po	At	Rn
Fr	Ra	Ac-Lr**	Rf	Db	Sg	Bh	Hs	Mt									

*Group of lanthanide	La 11.0	Ce 12.9	Pr 19.2	Nd 17.6	Pm	Sm 59.1	Eu 395	Gd 46.6	Tb 297	Dy 59.6	Ho 226	Er 48.7	Tm 649	Yb 125	Lu 896
**Group of actinide	Ac	Th 74.9	Pa	U 90.7	Np	Pu	Am	Cm	Bk	Cf	Es	Fm	Md	No	Lr

FIGURE 8.3 Cradle-to-gate GWP of various elements compared on a per kilogram basis. Data are representative of metal end-uses in 2008 using 2006 to 2010 price averages to allocate environmental impacts. Please see Nuss and Eckelman[49] for more details and additional effect categories. (From Nuss, P., Eckelman, M.J., Life cycle assessment of metals: A scientific synthesis. *PLoS ONE*, 9, e101298, 2014. Copyright: © 2014 Nuss, Eckelman. This is an open-access article distributed under the terms of the Creative Commons Attribution License, which permits unrestricted use, distribution, and reproduction in any medium, provided the original author and source are credited.)

metal use, or issues related to the vulnerability to supply restriction (e.g., competing end-use applications and metal substitution). These issues are important because many of the transition metals used in today's catalysts may not be available in quantities that guarantee sufficient supplies in the coming decades. Issues of resource availability are covered in depth in criticality assessments,[25–27] with outcomes for some of the metals used in sustainable metal catalysis (see above) reported at the global level and for the United States in Figure 8.4. The issue of critical materials in catalysts is further discussed in a study by the US National Research Council.[51]

Key point: Resource criticality assessments enable issues of resource concerns (of geological, geopolitical, and regulatory nature) to be identified and provide information about possible options to reduce resource constraints, for example, via substitution or alternative sourcing.

Figure 8.4 presents four metals of potential use in sustainable catalysis and their supply risk (SR) and vulnerability to supply restriction (VSR) scores. Iron and copper display low to moderate SR in the long-term VSR and medium VSR. Manganese indicates potential VSR issues, mostly related to issues of substitutability in some of its end-uses (e.g., steel making).[53] Zinc shows a slightly elevated SR (global level and national level) because its depletion time (i.e., the time until the geological reserve

Lower risk — Higher risk

Color key

Metal	Iron (Fe)	Manganese (Mn)	Copper (Cu)	Zinc (Zn)
SR (Global)[a] (0–100 axis)	0	2	5	41
SR (USA)[1] (0–100 axis)	45	61	52	57
VSR (Global)[2] (0–100 axis)	58	66	53	52
VSR (USA)[2b] (0–100 axis)	48	73	54	51
Reference	Nuss et al.[53]	Nuss et al.[53]	Nassar et al.[54]	Harper et al.[55]

FIGURE 8.4 Criticality results reported in the literature for some metals of interest in sustainable catalysis. The criticality methodology applied to derive these is given in Graedel et al.[27] All results are converted to a 0 to 100 (low–high) gray scale bar. [a]SR at the global scale consists of depletion time and companion metal fraction. At the national scale (United States), additional social and regulatory and geopolitical aspects of producing countries such as the policy potential index, human development index, political stability, and global supply concentration are included.[27] All scores are transformed to a 0 to 100 scale with 0 representing the lowest and 100 the highest SR. [b]VSR looks at issues such as importance, substitutability, and susceptibility of the metal under investigation.[27] All scores are transformed to a 0 to 100 scale with 0 representing the lowest and 100 the highest VSR.

base is exhausted taking into account current production figures, in-use lifetimes, and end-of-life recycling rates) is shorter than for any of the other elements in Figure 8.4.[55] Overall, any of these elements might be used in catalysis so long as a watchful eye is kept on the zinc and manganese situation.

8.3 LCA EXAMPLE: BIOBASED ACRYLIC ACID

As mentioned throughout this chapter, LCA is a useful tool to examine the environmental impacts of existing chemical routes as well as of processes still in their early stages of development. Here, we use literature data to investigate the life cycle–wide GHG emissions of biobased acrylic acid production and illustrate potential "hot spots" along the production chain. This process has not yet been developed commercially and therefore process data are scarce. However, the goal of this LCA is to illustrate how a streamlined LCA model can be developed even at an early stage of process development using best available literature data to obtain interesting insights, for example, into major contributors to environmental impacts. Such a basic LCA model, once developed, can be updated as better process data become available. It may be used internally within a company or research group when investigating the environmental performance of a novel synthesis route. Given the data limitations, we only focus on GWP as an effect category (in a full LCA, other impact categories should also be included).

Acrylic acid (2-propenoic acid) is commercially produced from petrochemicals, with the majority of it being generated by partial oxidation of propylene. Yields range between 50% and 60% for the single-step process and up to 90% for the two-step process via acrolein.[56] The major use of acrylic acid, and its derivatives, is in polymeric flocculants, dispersants, coatings, paints, adhesives, and binders for leather and textiles.[56] Currently, a variety of alternative routes from biomass to acrylic acid are under investigation. The biotechnological production from either 3-hydroxypropionic acid (3-HPA) or lactic acid is explored in the United States.[57] A number of companies and academic research groups are pursuing routes toward acrylic acid utilizing renewable feedstock. In a joint project, HTE (High Throughput Experimentation Company) and Arkema are working on the development of suitable catalysts for the conversion of biodiesel-derived glycerol to acrolein and acrylic acid.* The companies Cargill and Novozymes announced a joint $1.5 million agreement funded by the US Department of Energy to investigate technologies for the production of acrylic acid via 3-HPA from renewable biomass.† 3-HPA could act as precursor for acrylate production. Different hypothetical pathways for converting sugars into acrylate are summarized in the literature.[56] Each route may have particular advantages in terms of yield, productivity, ease of product separation, economy, and environmental impact. The most direct route is via lactate.

* See news article on http://biopol.free.fr/?p=597 (accessed March 17, 2010).

† See *Biomass Magazine* article on http://www.biomassmagazine.com/article.jsp?article_id=1407 (accessed March 17, 2010).

8.3.1 Life Cycle Inventory Data

The conversion process studied here is based on a hypothetical process flow diagram by Straathof et al.[56] In their study, a 100,000 tons/year fermentation process for acrylic acid production, including product recovery, was conceptually designed based on the assumption that an efficient host organism for acrylic acid production could be developed. It is assumed that conversion of the sugars toward acrylate can be achieved in a single step. A possible route that could be undertaken in this processing facility would be the fermentation of sugars through *S. cerevisiae* via pyruvate and lactate toward acrylic acid.[56,58] The production system considered in this study consists of (1) glucose production via corn wet milling, (2) fermentation in which the microorganism produces acrylic acid from fermentable sugars, (3) disk stack centrifuge for cell removal, (4) filtration system consisting of microfiltration and ultrafiltration, (5) liquid-liquid extraction of the product using an organic solvent, and (6) distillation of the solvent to obtain pure acrylic acid as the product.

Glucose production is based on a combination of data describing corn production in the United States (including farm equipment use)[59,60] and glucose production via corn wet milling[61–64] as described elsewhere.[65] The transportation distance and price data for allocation of environmental burden between glucose, corn meal and feed, and corn oil are based on the LCA model of Nuss and Gardner.[65] Data for the fermentation process is based on a yield of 0.72 g acrylic acid per gram glucose and a broth concentration of 50 g/L[66] to calculate the amount of fermentable sugars required and waste water produced (assuming no recirculation). Due to a lack of data on the detailed composition of wastewater generated, we use typical numbers from commercial polylactic acid (PLA) production according to the ecoinvent database[67,68] and Vink et al.[69,70] Approximately 0.10 g dry cells per gram of glucose are produced[58] and removed via filtration. Allocation of environmental burdens is based on the dry weight of product outputs. We do not include biogenic CO_2 emissions, for example, from cellular respiration during fermentation, and the uptake of biogenic CO_2 in the final product's effects on GWP. A liquid–liquid extraction process was chosen for the separation of acrylic acid from the mixture formed after fermentation and centrifugation. Alvarez et al.[71] found diisopropyl ether to be the most suitable solvent for acrylic acid from a water mixture. A diisopropyl ether requirement of 7.8 kg per kilogram of acrylic acid[71] and a solvent recovery rate of 0.95 kg per kilogram of waste solvent[72] are used in this assessment. Because no LCI data on diisopropyl ether are available in commercial LCI databases,[59] data for isopropyl acetate are used as a proxy.* Waste solvent not recovered is assumed to be treated in a hazardous waste incinerator.[73] The energy requirements of 3 kWh/m^3 feed for agitation and aeration and 7 kWh/m^3 feed for the centrifuge are taken from Patel et al.[74] Energy requirements for distillation (1.21 kg steam and 0.03 kWh electricity per kg waste solvent) and nitrogen requirements (0.24×10^{-3} Nm^3 per kg waste solvent) are obtained from Capello et al.[75] No data were available on the amount of sodium carbonate (used to adjust the pH) and the upstream environmental burden of biocatalyst

* Alvarez et al. state that, even though in their study, diisopropyl ether was used as a solvent, previous studies indicated the possibility of also using isopropyl acetate as a solvent for acrylic acid recovery.

production, and they are therefore excluded from the analysis. No infrastructure processes are included in this assessment. Electricity mixes of the background system represent the United States and all inputs and waste treatment processes are linked to respective unit processes from the US EI database.* The functional unit for this study is 1 kg of acrylic acid at the factory gate and the results are compared with conventional acrylic acid production using data from Althaus et al.[67] SimaPro8 LCA software is used to create the LCA model and effects on GWP are calculated using the IPCC 2013 GWP LCIA method with a 100-year timeframe.[76]

8.3.2 Results

The Sankey diagram for the production of bioacrylic acid from corn feedstock is shown in Figure 8.5. In the figure, the width of the arrows indicates the contributions of each unit process to total GWP. Unit processes located upstream or downstream of biobased acrylic acid production are shown as separate boxes in Figure 8.5 and represent conversion processes contributing to varying degree to GWP.

Key point: LCA allows a contribution analysis already in the design phase of a new chemical route or catalytic process. Based on this information, approaches can be developed to improve life cycle–wide performance.

Producing 1 kg of acrylic acid from corn feedstock is estimated to result in a GWP of approximately 4.9 kg CO_2-eq. The largest contributors to this are steam consumption during the distillation process (32% of total GWP) and corn wet milling during which glucose is produced (30%). This is mostly due to a combination of electricity, natural gas, and heavy fuel oil consumption associated with both processes. GHG emissions during corn production also contribute to the effects of corn wet milling. In addition, the use and subsequent disposal of solvents results in 22% and 10% of overall GWP, respectively. On-site electricity demand during fermentation contributes to 5%, only a small share to overall GWP. This compares with an estimated 2.4 kg CO_2-eq per kilogram of fossil-based acrylic acid (produced via a two-step oxidation process from propylene).[67]

However, large uncertainties are associated with many of the input parameters used in this streamlined LCA model. For example, the solvent/acrylic acid ratio and solvent recovery rate are based on simulation or laboratory-scale data and may not represent a commercial process in the future. The amount of solvent eventually required for the extraction process will vary with reaction conditions and the design of the fermenter (continuous or batch process, total throughput, type of waste solvent, size of the distillation equipment, etc.). For example, by decreasing the amount of waste solvent from 7.78 kg to 5 kg per kilogram of acrylic acid, the GWP would decrease to 4.3 kg CO_2-eq per kilogram of acrylic acid. Using a different solvent with a lower carbon footprint would further reduce these effects. Furthermore, steam generation is another major contributor to GHG emissions in this LCA. The amount required depends, among others, on the enthalpy of vaporization of the solvent and

* Using Ecoinvent 2.2.[59] with United States electricity mixes (www.earthshift.com/software/USEI-database).

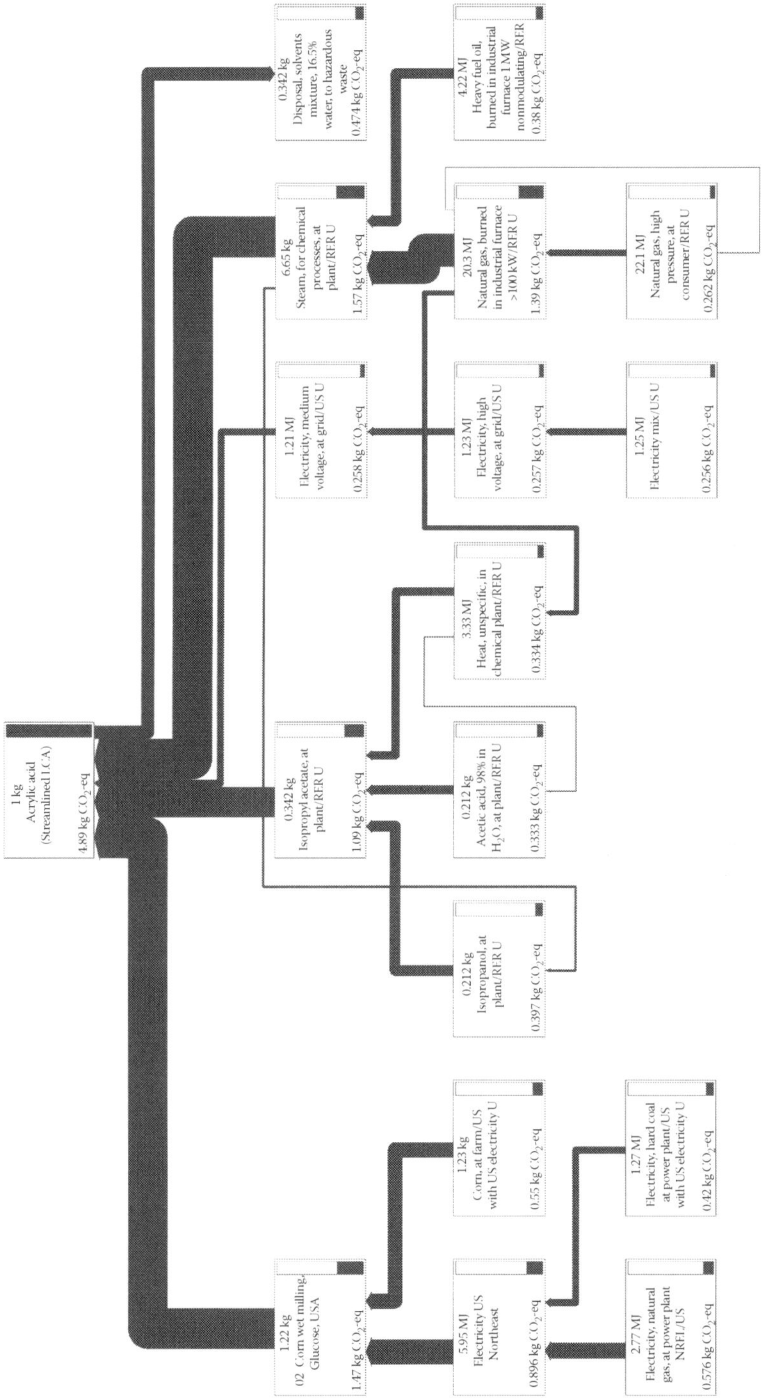

FIGURE 8.5 Sankey diagram showing the GWP associated with 1 kg corn-based acrylic acid production. Each box represents a unit process along the production chain of acrylic acid production and the width of the arrows indicates the contributions of each unit process to total GWP. For example, glucose provision via corn wet milling shown on the left hand side of the Sankey diagram contributes 1.47 kg CO_2-eq (~30% of total GWP).

may be significantly lower depending on the final solvent used in a commercial process. For example, cutting steam inputs by half would lower GWP to 4.15 kg CO_2-eq per kilogram of acrylic acid. Finally, switching to an alternative feedstock for fermentable sugar production (e.g., sugars obtained from lignocellulose feedstock) may help to further reduce environmental impacts.

Nevertheless, despite the issues of data availability discussed previously, this simple exercise demonstrates the capability of LCA to point toward potential "hot spots" in the life cycle of a chemical process, and it shows how various process settings may be examined early on in the design of a new chemical route. Furthermore, creating a first LCA model based on literature data usually takes no more than a few hours to perhaps a few days. If done in conjunction with process development in the laboratory, it can help the chemist to make more informed decisions and improve the life cycle–wide environmental performance of chemical synthesis.

8.4 CONCLUSIONS

Catalysis is an enabling technology that can help promote significant energy savings in the chemical industry and provide products of crucial importance for sustainable development in the future. However, several studies within the area of green chemistry and sustainable catalysis promote new techniques and products as "sustainable" or "more environmentally benign" without taking a life cycle–wide perspective and quantifying their benefits and tradeoffs. Life cycle approaches such as LCA and MFA can be used (oftentimes even during process development) to quantify environmental impacts and resource flows, and thereby help to elucidate improvement options and avoid shifting of (environmental) burdens to other parts of the life cycle. Resource criticality assessments can provide additional information with regard to possible resource constraints. It is hoped that these tools and approaches find increasing application in catalyst and chemical process design to help avoid unintended consequences and decrease our environmental impact on the earth.

Author's Note: Philip Nuss can be contacted at philip@nuss.me.

REFERENCES

1. Hoekstra, A. Y., Wiedmann, T. O. Humanity's unsustainable environmental footprint. *Science* 2014, 344, 1114–1117.
2. Bringezu, S., Schütz, H., Moll, S. Rationale for and interpretation of economy-wide materials flow analysis and derived indicators. *J. Ind. Ecol.* 2003, 7, 43–64.
3. Lettenmeier, M., Rohn, H., Liedtke, C., Schmidt-Bleek, F. *Resource Productivity in 7 Steps: How to Develop Eco-Innovative Products and Services and Improve Their Material Footprint*; Wuppertal Spezial 41; Wuppertal Institut für Klima, Umwelt, Energie GmbH: Wuppertal, Germany, 2009.
4. IEA. *Technology Roadmap: Energy and GHG Reductions in the Chemical Industry via Catalytic Processes*; International Energy Agency (IEA): Paris, France, 2013.
5. NRC. *Catalytic Process Technology*; National Research Council (NRC), National Academy Press: Washington, DC, 2000.

6. IHS. *Catalysts: Petroleum and Chemical Process*; IHS Chemical: Englewood, CO, 2013.
7. Ma, Z., Zaera, F. Heterogeneous catalysis by metals. In: *Encyclopedia of Inorganic and Bioinorganic Chemistry*; Scott, R. A., Ed., John Wiley & Sons, Ltd.: New York, 2006.
8. Howard, P., Morris, G., Sunley, G. Chapter 1—Introduction: Catalysis in the chemical industry. In: *Metal-Catalysis in Industrial Organic Processes*; Chiusoli, G. P., Maitlis, P. M., Eds., The Royal Society of Chemistry: London, UK, 2006; 1–22.
9. Armor, J. N. A history of industrial catalysis. *Catal. Today* 2011, 163, 3–9.
10. NRC. *Catalysis for Energy: Fundamental Science and Long-Term Impacts of the U.S. Department of Energy Basic Energy Science Catalysis Science Program*; National Research Council (NRC), National Academy Press: Washington, DC, 2009.
11. Plietker, B. Sustainability in catalysis—Concept or contradiction? *Synlett* 2010, 2049–2058.
12. Centi, G., Perathoner, S. Catalysis: Role and challenges for a sustainable energy. *Top. Catal.* 2009, 52, 948–961.
13. Kaneda, K., Mizugaki, T., Mitsudome, T. Green catalysis. In: *Encyclopedia of Catalysis*; John Wiley & Sons, Inc.: New York, 2002.
14. Centi, G., Perathoner, S. Catalysis and sustainable (green) chemistry. *Catal. Today* 2003, 77, 287–297.
15. Brundtland Commission. *Our Common Future*; Report of the World Commission on Environment and Development, Oxford University Press: Oxford, 1987.
16. Christensen, C. H. Catalysis for sustainability. *Top. Catal.* 2009, 52, 205.
17. Mebratu, D. Sustainability and sustainable development: Historical and conceptual review. *Environ. Impact Assess. Rev.* 1998, 18, 493–520.
18. Searchinger, T., Heimlich, R., Houghton, R. A., Dong, F., Elobeid, A., Fabiosa, J., Tokgoz, S., Hayes, D., Yu, T.-H. Use of U.S. croplands for biofuels increases greenhouse gases through emissions from land-use change. *Science* 2008, 319, 1238–1240.
19. Brunner, P. H., Rechberger, H. *Practical Handbook of Material Flow Analysis*; CRC Press: Boca Raton, FL, 2004.
20. ISO. *Environmental Management—Life Cycle Assessment—Principles and Framework, ISO14040*; ISO: Geneva, Switzerland, 2006.
21. ISO. *Environmental Management—Life Cycle Assessment—Requirements and Guidelines, ISO 14044*; ISO: Geneva, Switzerland, 2006.
22. Suh, S. *Handbook of Input–Output Economics in Industrial Ecology*; Springer: Dordrecht, the Netherlands, 2009.
23. Hunkeler, D., Lichtenvort, K., Rebitzer, G. *Environmental Life Cycle Costing*; CRC Press: New York, 2008.
24. UNEP. *Guidelines for Social Life Cycle Assessment of Products*; United Nations Environment Programme (UNEP): Paris, 2009.
25. EC. *Report on Critical Raw Materials for the EU*; Report of the Ad-hoc Working Group on defining critical raw materials; European Commission (EC): Brussels, Belgium, 2014.
26. NRC. *Minerals, Critical Minerals, and the U.S. Economy*; Committee on Critical Mineral Impacts of the U.S. Economy, Committee on Earth Resources, National Research Council. The National Academies Press: Washington, DC, 2008.
27. Graedel, T. E., Barr, R., Chandler, C., Chase, T., Choi, J., Christoffersen, L., Friedlander, E. et al. Methodology of metal criticality determination. *Environ. Sci. Technol.* 2012, 46, 1063–1070.
28. Wrisberg, N., Haes, H. A. U. D. *Analytical Tools for Environmental Design and Management in a Systems Perspective: The Combined Use of Analytical Tools*; Springer: Dordrecht, the Netherlands, 2002.
29. Klee, R., Graedel, T. E. Elemental cycles: A status report on human or natural dominance. *Annu. Rev. Env. Resour.* 2004, 29, 69–107.

30. Ayres, R., Simonis, U., Eds. *Industrial Metabolism: Restructuring for Sustainable Development*; United Nations University (UNU) Press: Tokyo, Japan, 1994.
31. Müller, D. B., Wang, T., Duval, B., Graedel, T. E. Exploring the engine of anthropogenic iron cycles. *Proc. Natl. Acad. Sci. U.S.A.* 2006, 103, 16111–16116.
32. Rudnick, R. L., Gao, S. 3.01—Composition of the continental crust. In *Treatise on Geochemistry*; Holland, H. D., Turekian, K. K., Eds., Pergamon: Oxford, 2003; 1–64.
33. Classen, M., Althaus, H.-J., Blaser, S., Scharnhorst, W., Tuchschmidt, M., Jungbluth, N., Faist-Emmenegger, M. *Life Cycle Inventories of Metals, Data v2.0*; Ecoinvent Report No. 10; Ecoinvent Centre, ETh Zurich: Dübendorf, CH, 2009.
34. Nassar, N. Chapter 7, Anthropospheric losses of platinum group elements. In: *Element Recovery and Sustainability*; Hunt, A. J., Ed., Cambridge, UK, 2013.
35. Saurat, M., Bringezu, S. Platinum group metal flows of Europe, Part 1. *J. Ind. Ecol.* 2008, 12, 754–767.
36. Ritthoff, M., Rohn, H., Liedtke, C. *Calculating MIPS—Resource Productivity of Products and Services*; Wuppertal Spezial 27e; Wuppertal Institut für Klima, Umwelt, Energie GmbH: Wuppertal, Germany, 2002; 52.
37. Swarr, T. E., Hunkeler, D., Klöpffer, W., Pesonen, H.-L., Ciroth, A., Brent, A. C., Pagan, R. Environmental life-cycle costing: A code of practice. *Int. J. Life Cycle Assess.* 2011, 16, 389–391.
38. UNEP. *Towards a Life Cycle Sustainability Assessment: Making Informed Choices on Products*; DTI/1412/PA; United Nations Environment Programme (UNEP): Paris, 2011.
39. Hellweg, S., Canals, L. M. I. Emerging approaches, challenges and opportunities in life cycle assessment. *Science* 2014, 344, 1109–1113.
40. Baumann, H., Tillman, A.-M. *The Hitch Hiker's Guide to LCA: An Orientation in Life Cycle Assessment Methodology and Application*; Studentlitteratur: Lund, Sweden, 2004.
41. Curran, M. A. *Life Cycle Assessment Handbook: A Guide for Environmentally Sustainable Products*, 1st ed.; Wiley-Scrivener: Hoboken, NJ, 2012.
42. Holman, P. A., Shonnard, D. R., Holles, J. H. Using life cycle assessment to guide catalysis research. *Ind. Eng. Chem. Res.* 2009, 48, 6668–6674.
43. Griffiths, O. G., Owen, R. E., O'Byrne, J. P., Mattia, D., Jones, M. D., McManus, M. C. Using life cycle assessment to measure the environmental performance of catalysts and directing research in the conversion of CO_2 into commodity chemicals: A look at the potential for fuels from "thin-air." *RSC Adv.* 2013, 3, 12244–12254.
44. Ravindra, P., Saralan, S., Abdulla, R. LCA studies for alkaline and enzyme catalyzed biodiesel production from palm oil. *Adv. Biol. Chem.* 2012, 2, 341–352.
45. Nexant. *PERP Program—Acrylic Acid New Report Alerts*; Nexant Chem Systems, 2006.
46. Enthaler, S., Junge, K., Beller, M. Sustainable metal catalysis with iron: From rust to a rising star? *Angew. Chem. Int. Ed.* 2008, 47, 3317–3321.
47. Marciniec, B. Hydrosilylation of alkenes and their derivatives. In: *Hydrosilylation*; Marciniec, B., Ed., Advances in Silicon Science; Springer: Netherlands, 2009; 3–51.
48. Tondreau, A. M., Atienza, C. C. H., Weller, K. J., Nye, S. A., Lewis, K. M., Delis, J. G. P., Chirik, P. J. Iron catalysts for selective anti-Markovnikov alkene hydrosilylation using tertiary silanes. *Science* 2012, 335, 567–570.
49. Nuss, P., Eckelman, M. J. Life cycle assessment of metals: A scientific synthesis. *PLoS One* 2014, 9, e101298.
50. Armor, J. N. So you think you may have a better process: How can you define the value? *Catal. Today* 2011, 178, 8–11.
51. U.S. National Research Council. *The Role of the Chemical Sciences in Finding Alternatives to Critical Resources: A Workshop Summary*; National Research Council (NRC), National Academy Press: Washington, DC, 2012.
52. Holwell, A. J. Global Release Liner Industry Conference 2008. *Platin. Met. Rev.* 2008, 52, 243–246.

53. Nuss, P., Harper, E. M., Nassar, N. T., Reck, B. K., Graedel, T. E. Criticality of iron and its principal alloying elements. *Environ. Sci. Technol.* 2014, 48, 4171–4177.
54. Nassar, N., Barr, R., Browning, M., Diao, Z., Friedlander, E., Harper, E. M., Henly, C. et al. Criticality of the geological copper family. *Environ. Sci. Technol.* 2012, 46, 1071–1078.
55. Harper, E. M., Kavlak, G., Burmeister, L., Eckelman, M. J., Erbis, S., Sebastian Espinoza, V., Nuss, P. et al. Criticality of the geological zinc, tin, and lead family. *J. Ind. Ecol.* 2014. (http://onlinelibrary.wiley.com.doi/10.1111/jiec.12213/abstract)
56. Straathof, A. J. J., Sie, S., Franco, T. T., van der Wielen, L. A. M. Feasibility of acrylic acid production by fermentation. *Appl. Microbiol. Biotechnol.* 2005, 67, 727–734.
57. Haveren, J. V., Scott, E. L., Sanders, J. Bulk chemicals from biomass. *Biofuels Bioprod. Biorefin.* 2008, 2, 41–57.
58. Lunelli, B., Duarte, E., Vasco de Toledo, E., Wolf Maciel, M., Maciel Filho, R. A new process for acrylic acid synthesis by fermentative process. *Appl. Biochem. Biotechnol.* 2007, 137–140, 487–499.
59. Ecoinvent. *Ecoinvent Life Cycle Inventory database v2.2*; Swiss Centre for Life Cycle Inventories: Dübendorf, Switzerland, 2010.
60. Jungbluth, N., Chudacoff, M., Dauriat, A., Dinkel, F., Doka, G., Emmenegger, M. F., Gnansounou, E. et al. *Life Cycle Inventories of Bioenergy*; Ecoinvent Report. No. 17; Swiss Centre for Life Cycle Inventories: Dübendorf, CH, 2007.
61. Akiyama, M., Tsuge, T., Doi, Y. Environmental life cycle comparison of polyhydroxyalkanoates produced from renewable carbon resources by bacterial fermentation. *Polym. Degrad. Stab.* 2003, 80, 183–194.
62. Gerngross, T. U. Can biotechnology move us towards a sustainable society? *Nat. Biotechnol.* 1999, 17, 541–544.
63. Khoo, H. H., Tan, R. B. H., Chng, K. W. L. Environmental impacts of conventional plastic and bio-based carrier bags. *Int. J. Life Cycle Assess.* 2010, 15, 284–293.
64. U.S. EPA. *Chapter 9: Food and Agricultural Industries*, AP 42, 5th ed., vol. I; United States Environmental Protection Agency (USEPA), Office of Air Quality Planning and Standards, 2011.
65. Nuss, P., Gardner, K. H. Attributional life cycle assessment (ALCA) of polyitaconic acid production from northeast US softwood biomass. *Int. J. Life Cycle Assess.* 2013, 18 (3): 603–612.
66. Hermann, B., Patel, M. Today's and tomorrow's bio-based bulk chemicals from white biotechnology. *Appl. Biochem. Biotechnol.* 2007, 136, 361–388.
67. Althaus, H.-J., Hischier, R., Osses, M., Primas, A., Hellweg, S., Jungbluth, N., Chudacoff, M. *Life Cycle Inventories of Chemicals Data v2.0*; Ecoinvent Report No. 8; Ecoinvent Centre, ETH Zurich: Dübendorf, CH, 2007.
68. Althaus, H.-J., Werner, F., Stettler, C. *Life Cycle Inventories of Renewable Materials Data v2.0*; Ecoinvent Report No. 21; Ecoinvent Centre, ETh Zurich: Dübendorf, CH, 2007.
69. Vink, E. T. H., Rábago, K. R., Glassner, D. A., Gruber, P. R. Applications of life cycle assessment to NatureWorks™ polylactide (PLA) production. *Polym. Degrad. Stab.* 2003, 80, 403–419.
70. Vink, E. T. H., Glassner, D., Kolstad, J., Wooley, R., O'Connor, R. The eco-profiles for current and near-future NatureWorks® polylactide (PLA) production. *Ind. Biotechnol.* 2007, 3, 58–81.
71. Alvarez, M., Moraes, E., Machado, A., Filho, R., Wolf-Maciel, M. Evaluation of liquid–liquid extraction process for separating acrylic acid produced from renewable sugars. *Appl. Biochem. Biotechnol.* 2007, 137–140, 451–461.
72. Geisler, G., Hofstetter, T., Hungerbühler, K. Production of fine and specialty chemicals: Procedure for the estimation of LCIs. *Int. J. Life Cycle Assess.* 2004, 9, 101–113.

73. Doka, G. *Life Cycle Inventories of Waste Treatment Services—Part II "Waste Incineration"*; Ecoinvent Report No. 13; Swiss Centre for Life Cycle Inventories: St. Gallen, Switzerland, 2009.
74. Patel, M., Crank, M., Dornburg, V., Hermann, B. G., Roes, L., Hüsing, B., Overbeek, L., Terragni, F., Recchia, E. *Medium and Long-Term Opportunities and Risks of the Biotechnological Production of Bulk Chemicals from Renewable Resources—The Potential of White Biotechnology*. The BREW Project; Technical Report; Utrecht University: Utrecht, the Netherlands, 2006.
75. Capello, C., Hellweg, S., Badertscher, B., Hungerbühler, K. Life cycle inventory of waste solvent distillation: Statistical analysis of empirical data. *Environ. Sci. Technol.* 2005, 39, 5885–5892.
76. Goedkoop, M., Oele, M., de Schryver, A., Vieira, M. *SimaPro Database Manual—Methods Library*. Report version 2.2; PRé Consultants: Netherlands, 2008.

Index

Page numbers followed by f and t indicate figures and tables, respectively.

H

I

J

K

L

O

P

Made in the USA
Las Vegas, NV
19 September 2026

dac84e84-50e0-4cca-8492-e87f2b8ca54cR01